"An eminently important book that illustrates the expanded scope of psychoanalytic thought and action when nature is recognised as a primary psychic object alongside the primary parental objects of humans. With an integrative concept of climate in mind, we psychoanalysts can no longer keep separated the intimate familial climate from the physical climate and the wider social climate. This has fundamental consequences for the form and content of psychoanalytic interpretations: the examination of the unconscious and conscious meaning of all climatic themes requires our attention, including the concern for the preservation of nature and its representation in the individual psyche. This stirring, lucid, and evocative book emphasises the social responsibility of psychoanalysis, and that deserves many readers because it convincingly conveys the urgency of psychoanalytically contributing to the transformation of the currently widespread culture of uncare into a more lively culture of care."

Dr. med. Heribert Blass,
President International Psychoanalytical Association

"At last, as this volume reveals, psychoanalysts across the globe are waking up to the promptings of nature and climate. Not just an ethical imperative, its contributors also demonstrate how this shifting orientation enriches psychoanalytic theory and practice, not least by deepening our understanding of key aspects of the human condition such as omnipotence, splitting, suffering, love and reparation."

Professor Emeritus Paul Hoggett,
Co-Founder Climate Psychology Alliance

"As a scientist and ecologist watching what is happening to our natural environments, I have always wondered why so many humans show such little respect for Nature. I am now beginning to understand why. This collection of essays provides vital insights and clarifications into the core causes of human behavior with regards to Nature and the Earth. The contributors provide an impressive array of perspectives to expand our comprehension, and hopefully to provide remedies, to the growing disjunct in the 21st Century between our human interior world and the external reality of Nature, climate, and planet. It is a must read for psychoanalysts, scientists, and even the general public, who are struggling to understand this disjunction."

W. John Kress, PhD, *Distinguished Scientist and Curator Emeritus,*
National Museum of Natural History Smithsonian Institution

With Climate in Mind

This groundbreaking book enriches and expands psychoanalytic theory and method as it applies to the climate.

It embeds psychoanalysis within environmental and cultural awareness; restores the importance Freud placed on external reality and its mental representations; introduces an integrative concept of climate; and, with its attention to clinical detail, offers stepping stones for practitioners seeking to understand clinical material in which phantasies involving nature, culture, and family are intertwined. Presented in four parts – Clinical, Theory, Nature, and Research – its authors are psychoanalysts from across the world.

With Climate in Mind is essential reading for practising and training psychoanalysts, for those in the psychotherapy profession, and for other professionals engaged with what climate breakdown in a culture of carelessness means today.

Sally Weintrobe is Fellow of the British Psychoanalytical Society and has chaired the IPA Climate Committee. She won the IPA's 2021 Community Award for her work on climate. *Engaging with Climate Change* (Routledge), which she edited, was short-listed in 2014 for the Gradiva Award for its contribution to psychoanalysis.

Lynne Zeavin is Training and Supervising Analyst of the New York Psychoanalytic Society and Institute and an associate editor of *JAPA*, the *Journal of the American Psychoanalytic Association*. Among her publications, she co-edited, with Don Moss, *Hating, Abhorring and Wishing to Destroy* (Routledge).

IPA in the Community
Series Editor Harvey Schwartz

Applying Psychoanalysis in Medical Care
Edited by Harvey Schwartz

Trauma, Flight and Migration
Psychoanalytic Perspectives
Edited by Vivienne Elton, Marianne Leuzinger-Bohleber,
Gertraud Schlesinger-Kipp and Vivian B Pender

Psychoanalytic Work in East Africa
Barbara Saegesser

Exploring Eating Disorders Through Psychoanalysis
Unravelling the Psyche
Humberto Lorenzo Persano

A Psychoanalytic Approach to Refugee Mental Health
Safe Harbor
Chrysi Giannoulaki

With Climate in Mind
Psychoanalysts on Climate Breakdown
Edited by Sally Weintrobe and Lynne Zeavin

For more information about this series, please visit: www.routledge.com/IPA-in-the-Community/book-series/IPAC

With Climate in Mind

Psychoanalysts on Climate Breakdown

Edited by Sally Weintrobe
and Lynne Zeavin

Routledge
Taylor & Francis Group

LONDON AND NEW YORK

Designed cover image: Getty | Ernst Haas

First published 2026
by Routledge
4 Park Square, Milton Park, Abingdon, Oxon OX14 4RN

and by Routledge
605 Third Avenue, New York, NY 10158

Routledge is an imprint of the Taylor & Francis Group, an informa business

British Library Cataloguing-in-Publication Data
A catalogue record for this book is available from the British Library

ISBN: 978-1-041-06339-1 (hbk)
ISBN: 978-1-041-06338-4 (pbk)
ISBN: 978-1-003-63495-9 (ebk)

DOI: 10.4324/9781003634959

Typeset in Palatino
by Apex CoVantage, LLC

We dedicate this book to our grandchildren, Ollie Humphreys, Jessica Humphreys, Lotte Rometsch, Lusha Rometsch, Roko Plevnik, Mila Plevnik, and Malachai Timothy Zeavin O'Brien

Contents

Acknowledgements

We wish to thank our fellow members of the IPA Climate Committee, namely Maria Luiza Gastal, Delaram Habibi-Kohlen, Mary-Anne Smith, Karyn Todes, Davide Rosso, Valerie Curen, and Pushpa Misra (a former member), who have met monthly over the last five years and generated a rare atmosphere of collegiality and intellectual sustenance and fortification in very dark times – beginning with the advent of COVID-19 through to the present. Many, but not all, of the chapters in the book come from members of this committee.

Many of the chapters in this volume are published here for the first time. We are very grateful to the authors for their engagement and fine work. Other chapters have appeared previously in journals or as book chapters. Lindsay L. Clarkson's chapter in this book is a revision of a chapter, "Trees and Other Psychoanalytic Matters", that was included in an anthology, *Hating, Abhorring and Wishing to Destroy*, edited by Don Moss and Lynne Zeavin, Routledge (2022). It is reproduced by permission of Taylor & Francis Group. We are grateful to the publisher for permission to reprint Bauriedl-Schmidt, C., Fellner, M., Krimmer, M., Wirth, H.-J. (2022), 'Was ist psychoanalytische Aufklärung heute? Eine Kultur der Fürsorge als Antwort auf die Verletzbarkeit des Subjekts im Angesicht der Klimakrise', originally published in *Psyche – Z. Psychoanal.*, 76.8, pp. 734–44. DOI: 10.21706/ps-76-8-734. Alfredo Lombardozzi's chapter is reprinted here with the permission of the author.

Our thanks to Delaram Habibi-Kohlen for editorial assistance and to Harvey Schwartz, the editor of this series for Routledge. We are indebted to Jessica Gossling and Alice Condé for their careful, consistent, and highly valued editorial assistance. Thank you, both of you.

And finally, we wish to thank Lez and Don for their support and engagement with these vital concerns.

Contributors

Christine Bauriedl-Schmidt, is a psychoanalyst and group therapist. She is the incoming Co-Chair of the German Society for Psychoanalysis, Psychotherapy, Psychosomatics and Depth Psychology (DGPT), where she also serves on the board. She is a lecturer, training analyst, and supervising analyst at the Munich Association for Psychoanalysis (MAP). She is an associate member of the *Deutsche Gesellschaft für Gruppenanalyse* (D3G) and a member of *Psychologists/Psychotherapists for Future* (Psy4F). Trained as a psychologist, she holds a doctorate in human biology (Dr. rer. biol. hum.) from the University of Erlangen–Nürnberg.

Lindsay L. Clarkson, a psychiatrist and psychoanalyst, is currently a supervising and training analyst emerita at the Washington Baltimore Center for Psychoanalysis and a member of the Humanities and Psychoanalysis Study Group at Dartmouth. She explores our relationship to the natural world in the hope that such enquiry might contribute to a more benign and protective response to the evolving environmental catastrophe. Dr. Clarkson combines a lifelong interest in natural history with an appreciation of Klein and Bion, as well as the contemporary aspects of their tradition and perspective. She has used clinical process, narrative and environmental literature, poetry, and biography to broaden the scope of psychoanalytic listening and developmental theory to include the role of the environment within our internal worlds. In addition to presenting on various aspects of this topic at meetings of APSA and the IPA, and as an invited speaker at society meetings, her essays have been published in four anthologies. Her most recent journal contribution, co-authored with Dr. Shelley Rockwell, appeared in *JAPA* 2024, "The Weight and Heft of the Natural World in Our Inner Selves".

Markus Fellner is a psychoanalyst working with children, adolescents, and adults, as well as a certified family therapist (DGSF). He serves as a member, lecturer, and training and supervision analyst of the Deutsche Gesellschaft für Psychoanalyse, Psychotherapie, Psychosomatik und Tiefenpsychologie (DGPT) and the Münchner Arbeitsgemeinschaft für Psychoanalyse (MAP). In addition to his clinical and teaching activities, he is actively engaged in professional networks that connect psychoanalysis with broader social and ecological concerns, including Psychologists/Psychotherapists for Future (Psy4F).

Maria Luiza Gastal is a psychoanalyst and an associate member of the Brasília Psychoanalysis Society (SPBsB). She is a professor at the Virginia Leone Bicudo Psychoanalysis Institute in Brasília and a member of the IPA Climate Committee. Trained as a biologist, she holds a PhD in ecology from the University of Brasília. She has served as a biodiversity consultant for the Brazilian Ministry of Environment and is a retired professor from the University of Brasília (UnB).

Delaram Habibi-Kohlen is a psychoanalyst and training analyst in private practice in Bergisch Gladbach, Germany. She is a member of the DPV, DGPT, and IPA. Her publications focus on societal and political issues – especially the denial of climate change – as well as microprocesses in transference and countertransference.

Monika Krimmer now works as a medical psychotherapist and psychoanalyst for individuals and groups in public practice and is also qualified anesthetist and pain specialist. She is a lecturer at several psychotherapeutic institutes as well as psychiatric hospitals.

Alfredo Lombardozzi is an individual and group psychoanalyst and a training analyst of the Italian Psychoanalytic Society (SPI) and the International Psychoanalytic Association (IPA). Trained as an anthropologist, he has taught psychoanalytic anthropology at the Universities of Chieti and L'Aquila. He was editor of the journal *Koinos – Gruppo e funzione analitica* (Koinos – Group and Analytical Function) and *the Rivista di Psicoanalisi* (Journal of Psychoanalysis). He has published many scientific articles on the relationship between psychoanalysis and anthropology and has edited several volumes on related topics. He is also the author of *Figure del dialogo tra antropologia e psicoanalisi* (Figures of Dialogue Between Anthropology and Psychoanalysis) (Borla, 2006); *L'imperfezione dell'identità. Riflessioni tra psicoanalisi e antropologia* (The Imperfection of Identity. Reflections between Psychoanalysis and Anthropology) (AlpesItalia, 2015); and *Culture di gruppo. Per un'antropologia del gruppo psicoanalitico* (Group Cultures. Towards an Anthropology of the Psychoanalytic Group) (AlpesItalia, 2021).

Luc Magnenat is a psychoanalyst in private practice in Geneva and a training member of the Swiss Psychoanalytical Society. He is interested in developments in ecological thinking, environmental sciences, and the impact of the environmental crisis on both our lives and psychoanalytic practice. In 2019, he edited *La crise environnementale sur le divan* (The Environmental Crisis on the Couch), a collective interdisciplinary work that brings together environmental scientists and psychoanalysts (published by In Press, Paris). He edited *Résister à la crise environnementale* (Resisting the Environmental Crisis) in 2025, also published by In Press, Paris. He is the laureate of the 2000 Germaine Guex Scientific Prize awarded by the Swiss Psychoanalytical Society, the 2018 Maurice Haber Prize awarded by the Belgian Review of Psychoanalysis, and the 2023 Climate Prize of the International Psychoanalytic Association.

Pushpa Misra is a training and supervising analyst, and a former President of Indian Psychoanalytical Society. After doing her Masters' in Psychology and Philosophy, Pushpa did her Ph.D. in Philosophy from the University of Rochester, New York. She did her postdoctoral work as a Fulbright fellow from the University of Pittsburgh, in 1993–1994. Her research work was published by Karnac Books, London, titled "Scientific Status of Psychoanalysis: Evidence and Confirmation".

Don Moss is a psychiatrist and psychoanalyst. He is the author of more than 60 articles and five books, most recently *Traumatizing Disorders of Everyday Life* (JAPA) and *Psychoanalysis in a Plague Year* (Routledge). He is the recipient of the Elisabeth Young-Bruehl prize for his work against prejudice and the Haskell Norman Prize for excellence in psychoanalysis. He is a founding member of Green Gang, a group of analysts and scientists dedicated to studying humans' relationship with the natural world. He serves on the editorial boards of IJP and JAPA and has been practising psychoanalytic work in New York for 45 years.

Jhelum Podder is a psychoanalyst, writer, and former psychology professor based in Kolkata, India. She trained with the Indian Psychoanalytical Society and is affiliated with the International Psychoanalytical Association.

Harold Searles was an influential American psychiatrist and psychoanalyst, best known for pioneering psychoanalytic treatment of schizophrenia. Educated at Cornell and Harvard Medical School, he trained at Chestnut Lodge and the Veterans Administration. Searles emphasized countertransference and the therapeutic relationship, shaping modern psychoanalysis with his profound insights into the complexities of human connection and mental illness.

Karyn Todes is a training psychoanalyst and a clinical and counselling psychologist in private practice in Sydney, Australia. She is a member of the Australian Psychoanalytic Society (APAS), the International Psychoanalytic Association (IPA), and the Australian Psychological Society (APS). She is a committee member on the IPA Climate Committee (since 2020) and the IPA Scientific Committee (since 2023), working across regions to advance psychoanalysis.

Sally Weintrobe, Fellow of the British Psychoanalytical Society, chairs the Climate Committee of the International Psychoanalytical Association, is a member of the Climate Psychology Alliance, and serves as a board member of the Ecopsychepedia, an online free resource on the mental-health impacts of global heating. She was one of the 31 Global Commissioners for the Cambridge Sustainability Report (2021). She has written and talked widely on psychological drivers of climate break-down. Her books on climate include *Engaging with Climate Change* (Rout-ledge and New Library of Psychoanalysis, 2021, editor) and *Psychological Roots of the Climate Crisis: Neoliberal Exceptionalism and the Culture of Uncare* (Bloomsbury, 2021, author).

Hans-Jürgen Wirth is a psychological psychotherapist, psychoanalyst, and psychoanalytic couple and family therapist in private practice. He is Pro-fessor of Sociology and Psychoanalytic Social Psychology at the Univer-sity of Frankfurt am Main. In addition to his academic and clinical work, he is the founder of Psychosozial-Verlag and serves as co-editor of the journals *psychosozial* and *Psychoanalytische Familientherapie*. His scholarly contributions span psychoanalysis, politics, and society, with selected publications including *9/11 as a Collective Trauma and Other Essays on Psy-choanalysis and Society* (2004), *Narcissism and Power* (5th ed., 2015), and *Gefühle machen Politik: Populismus, Ressentiments und die Chancen der Ver-letzlichkeit* (2022).

Lynne Zeavin is a clinical psychologist and psychoanalyst in full-time practice in New York City. She is a training and supervising analyst at the New York Psychoanalytic Society & Institute, where she chairs the curriculum. An associate editor at JAPA, she has authored papers exploring idealization, the status of the object, neutrality, interpretation, mourning, and various aspects of Kleinian theory. Dr. Zeavin supervises widely from a contemporary Kleinian perspective. A member of the IPA Work Group on Climate, she is co-founder of the Rita Frankiel Memorial Fellowship, funded by the Melanie Klein Trust, and a founder of Second Story, a non-institutional psychoanalytic space in New York City. She is co-editor, with Don Moss, of *Hating, Abhorring and Wishing to Destroy: Psychoanalytic Essays on the Contemporary Moment* (2021).

Introduction

Sally Weintrobe and Lynne Zeavin

Psychoanalysis has recently been undergoing seismic changes in theory, clinical technique, and the quality of its social commentary. We believe that the most significant driver of these changes is that, at this point in history, an increasing number of psychoanalysts are paying far greater attention to external reality than before. This focus on external reality is in some respects entirely new, for reasons that will be explored later – the main one being that external reality, in the shape of the physical climate (including the natural world) and the wider social climate, is rapidly changing. Although they are separate areas of study, it is increasingly clear that each climate profoundly affects the other, for good and ill. The rapid changes we are seeing in both climates are now so great that external reality can no longer be credibly ignored.

Of course, psychoanalysts have been interested in external reality ever since Freud included the study of reality as fundamental to understanding the mind. It is beyond our scope here to explore in detail why so many psychoanalysts in recent decades – and up to the present day – have treated large parts of external reality as outside of their field of study and have focused mainly on the immediate family and its climate. However, this has made it hard for colleagues to be taken seriously when they argue that the state of the climate in a wider sense is part of the study of mind.

The chapters in this remarkable book, which we are proud to edit, testify to the shifts that psychoanalysis is currently undergoing. The authors are all practising psychoanalysts, with many serving as members of the Climate Committee established by the International Psychoanalytical Association in 2019. Being based in countries across the world – and thus embedded in climates that differ physically, socially, and culturally – we believe enriches the fabric that these authors weave together immeasurably.

As part of this collection, we have included Harold Searles' 1972 pioneering paper, "Unconscious Processes in Relation to the Environmental Crisis" (Chapter 7), which has been largely ignored by the psychoanalytic community until recently. In it, he foresaw with striking accuracy where humanity would end up if omnipotent fantasy gained sufficient power to triumph over reality-based thinking – that is, the capacity to recognize real

DOI: 10.4324/9781003634959-1

limits – the "No of Nature", as Ro Randall put it[1] –and to observe those limits for life itself to endure. Searles draws out one of Freud's insights: phantasy, when empowered, can alter the material world. Searles' point was that omnipotent fantasy, when not adequately held in check by an awareness of limits, can alter material reality in harmful ways rather than serving the common good.

Attention to external reality has led to various shifts that are evident in this edited collection: a widening of what is included as falling within the psychoanalytic field of enquiry; a sharper interest in what might count as a psychoanalytic theory of ethics; greater attention to the ordinary, healthy mind alongside pathological states in order better to understand what supports human flourishing; and finally, a more humble tone. These signs of taking external reality seriously, we believe, have the potential to advance and invigorate psychoanalysis as a discipline and to strengthen its ties with, and relevance to, other disciplines.

Climate as Nature

These authors are porous to nature in the sense that they are open to taking it in, introjecting it, and allowing themselves to be affected by it in manifold ways: its beauty, its power to calm and heal, and its ability to provide sustenance, all of which support life, liveliness, and the development of mind. They are also porous to taking in and being affected by the extent of nature's current degradation. They are not estranged from nature, as so many find themselves when swayed unconsciously by what Sally Weintrobe has called the culture of uncare.[2] Such a culture, she argues, operates as a pathological organization that undermines efforts to connect and to make reparation. In many respects, the authors of the chapters in this volume come across as having managed to hold on to – or to have repaired – their relationship with their childhood selves, in awe of, and unambiguously attracted to and in love with, nature. This quality of fresh-eyed wonder at nature's beneficence (arguably forming part of what is termed the "good object") is particularly evident in the chapters by Lindsay L. Clarkson (Chapter 12), Pushpa Misra (Chapter 14), and Lynne Zeavin (Chapter 3), as well as in the people interviewed by Misra and Jhelum Podder in their research into feelings about nature in northern India (Chapter 16), and in many other chapters in this edited collection. It leads several authors to assert that nature is an object of primary identification.

The title of this book, *With Climate in Mind*, highlights the place of nature as *in the mind*. When viewed psychoanalytically, nature is a psychic object that has acquired representation as part of what Melanie Klein referred to as the representational world – the question being whether that representation is coloured by the reality principle or the pleasure principle. In addition to locating nature in and as part of the mind, many of these authors make the stronger claim that nature is a primary psychic object, alongside the primary parental objects.

There is a schism within psychoanalysis between those who do not accept nature as a primary psychic object and those who suggest that it is one – a new position put forward in various chapters of this book. The former group regularly tends to treat clinical material that contains references to nature as derivative or metaphorical, seeing it as standing in for, or being like, what is considered the "real" subject. Weintrobe (Chapter 4) parodies this stance with a scene from the film *Sex, Lies, and Videotape* (1989), in which the patient talks about her worry over all the waste and rubbish piling up in the world, and her therapist replies that she is talking about the state of her marriage. She may be. However, parody aside, there remains a widespread tendency to treat environmental imagery and fantasy as mere metaphor, rather than as something with the potential to be central to the session – its point of urgency.

Luc Magnenat (Chapter 15), by contrast, draws our attention to the real, palpable and now general worry about human-caused toxic waste, pollution and environmental damage, referring to this as a hyper-object. Mostly kept unconscious, the hyper-object now seems to be looming over all our lives. He also discusses the loss of what Joseph Sandler called our "background of safety" as humanity transitions from the stability of the Holocene to the Anthropocene era.[3] He extends Sandler's concept to include the holding-stable environment provided by flourishing life-support systems. The loss of safety brings with it profound distress, making it potentially harder to keep *climate in mind*. Donna Haraway called this "staying with the trouble".[4]

If the woman just quoted was voicing her concern about actual environmental waste, she may well have felt silenced, unheard, and not met by the interpretation she received. As many of these authors make clear by "showing their workings" when discussing their clinical material, disentangling the personal from more general strands (separating "I" from more collective "we" phantasies) and distinguishing between external and internal realities is complex work – work that is bypassed if the analyst's initial stance is to screen out or disavow (minimize) material concerning core areas of external reality, such as nature.

Many psychoanalysts suddenly became more porous to nature in their "on the couch" work with the arrival of the COVID-19 pandemic – a time when climate effects across the globe were also becoming too huge to ignore. Nature entered the consulting room in the form of the virus and as climate-related phenomena. One example is a patient in Sydney, Australia, arriving for a session during wildfires, coughing and spluttering before saying, "Give me a moment. The fire smoke is in my lungs" (personal communication from Karyn Todes. She describes in Chapter 2 how the wildfires affected her capacity to hold the setting in her mind with an aggressive, perverse patient).

The positioning of nature as beyond the psychoanalytic field of enquiry seems to indicate a profound estrangement from nature. We might speculate that this reflects a broader estrangement from nature in more

Westernized cultures. This issue is raised in various ways across many of the chapters. In *Living in Truth* (1987), Vaclav Havel describes small children as "pre-political" in their love of nature.[5] He meant they are not yet heavily influenced by a prevailing culture – especially one that encourages people to see nature as something to dominate, extract from, and control. To take nature and its current state seriously is to suffer in Wilfred Bion's sense: to be open to experience and to learn from it in ways that change you – sometimes leaving you feeling potentially overwhelmed, emotionally undone, grief-stricken, and fragmented as a result. Suffering in this sense also includes the experience of joy, delight, and wonder at nature, which can be so intense that it brings tears to one's eyes. For Bion, thinking is a feelingful affair.

In this edited collection, being porous to nature is reflected in a shift in tone. One way this is expressed is through the authors' openness to tolerating not knowing. Psychoanalysts are familiar with the importance of "negative capability" (enduring and suffering a state of not knowing). However, the porosity described here appears to be part of a wider, cross-disciplinary movement towards embracing a view of life, and the systems that support it, as more complex, interconnected, and entangled than we have ever imagined or can fully comprehend. While this has always been known – and indeed Freud pointed out that meaning is never finally unravelled – at this point in history, we are confronted anew with the extent of our ignorance because so much more is now understood about the complexity of life systems.

In *With Climate in Mind*, the authors show readers the changes and adjustments in theory, practice, and within themselves as they become more inclusive and more open to the true state of the climate and the natural environment. Through their struggles to incorporate emerging new facts from both the natural and human sciences, they provide helpful stepping stones for those of us trying to find our way in our psychoanalytic work today. For example, many of the authors preface their clinical material by explaining why they have expanded the boundaries of what they include as part of the psychoanalytic terrain, and how they have done so. Many describe how confronting climate reality has meant exposing themselves to a level of suffering that the ecologist Arne Naess characterized as the deep wounds felt upon gaining an ecological education. The fact that the authors are embedded in different geopolitical contexts, as well as in different traditions within psychoanalysis, provides a rich array of theoretical perspectives. For example, Chapter 8 by Christine Bauriedl-Schmidt, Markus Fellner, Monika Krimmer, and Hans-Jürgen Wirth is rooted in the psycho-social approach developed by the Frankfurt School; Chapter 5 by Alfredo Lombardozzi brings in the thinking of Italian colleagues; and Chapter 9 by Maria Luiza Gastal reveals a psychoanalytic perspective that has been enriched by contemporary Brazilian Indigenous thinkers such as Ailton Krenak.

This edited collection is full of examples showing how serious attention to external realities can open up both theoretical and clinical understanding.

Delaram Habibi-Kohlen (Chapter 1), for instance, writes about her patient's distress at seeing dead trees in the forest. She realizes that she too is deeply saddened by the growing number of trees dying in Germany as a result of global warming. She takes two interconnected steps, and one consequence is that she becomes filled with sadness. First, she does not reflexively assume that the material about the trees is "really" about something else. Secondly, she locates herself and her patient as living in the same world – a world that affects them both. This, at the very least, provides a framework for reflecting on what might belong to the patient, to the analyst, and what is likely to be widely shared by people within a given context. It enables the possibility of a more fine-grained clinical understanding. It also reflects an underlying ethical perspective that is not based on splitting into "us" and "them".

Climate is seen as affecting us all, even if unequally.

Climate as Culture

Many of the authors in this collection broaden their setting to include culture – as group pressure, as ideology or as hidden persuasion by vested interests. Don Moss (Chapter 6) discusses a kind of splitting that positions people hierarchically – as being above (and superior) or below (and inferior).

Calling this "vertical splitting", he exemplifies its presence – and the cruelty it unleashes – in the clinical situation he describes. Vertical splitting dehumanizes and ensures injustice, a point emphasized by Moss. He notes its powerful structuring presence within the Hebraic-Christian doctrine that man shall hold dominion over all he surveys. The Nigerian psychologist Bayo Akomolafe, who refers to himself as a "recovering psychologist" (meaning he is recovering from the fracturing mental effects of splitting driven by colonialism), conveys this dehumanization through the image of a slave ship split vertically down the middle to reveal white slave masters (note the gender bias) on the upper deck, black slaves on the deck below, and animals down in the hold.

Vertical splitting, of course, also restructures power relations within the self. As Klein noted, when we split, we also split the self. Bauriedl-Schmidt, Fellner, Krimmer, and Wirth take up this theme in Chapter 8, exploring how people come to suppress their lively, caring, truth-seeking part. Weintrobe has discussed this internal power relationship as being maintained through narcissistic entitlement, reinforced by culture and ideology.[6] Here, the more caring, mindful part is split off and consigned to an abject, non-entitled position. Gastal (Chapter 9) addresses the need for a "recovering psychoanalysis" in her exploration of how the colonial mindset infiltrated psychoanalytic theory, with its early presumption that the thinking of indigenous people was primitive, and its implicit racism. The authors of this book are all recovering psychoanalysts, in the sense discussed here, struggling to identify split states of mind – not least within themselves – and to reconnect with good objects that have been lost, demeaned, or damaged. All of this is essential to reparation – the ongoing depressive position, psychic work essential to life and to its flourishing.

Reparation

Reparation is Zeavin's central theme in Chapter 3, which focuses on a patient's struggle to repair her relationships. She explores how the capacity for reparation is essential to the ability to love, care, and feel concern. Building on this, we argue that reparation lies at the heart of the life instinct and is vital to the preservation of life.

Across *With Climate in Mind*, the authors draw attention to human cruelty and attacks on what supports life and vitality at various levels and in different spheres: within the psyche, in family groups, social settings, and society at large. Zeavin describes a clinical case in which her patient's excitement arising from cruel physical and verbal abuse prevents her from connecting with her numbed inner world, making it extremely difficult for her to repair her human relationships. Todes' patient (Chapter 2), a man who had experienced severe early trauma due to parental violence and neglect, also obstructed the process of repair through excited, sadistic attacks. Zeavin's patient appeared to attain greater mental stability – or "stillness", as Zeavin describes it – through reparative acts of caring for nature (her garden and her cat), which she found less threatening than repairing her human relationships. For Todes' patient, nature seemed primarily to consist of bad, threatening objects (wild attacking animals). His level of paranoia appeared to prevent him from turning to nature for reparation and solace.

The material from these patients confronts us not only with how vital ongoing reparation is (Zeavin's main point) but also with how difficult it can be to make repairs. It also confronts us with the irreparable. Todes' patient seemed to achieve some degree of calm (the excited beating subsided) in recognizing that he was damaged – a hollow man – a painful insight that can itself be seen as an act of mental repair. Reparation can restore hopefulness, a theme explored in detail by Luc Magnenat (Chapter 15), who draws on Jonathan Lear's concept of radical hope in the face of devastating destruction and loss. Radical hope is bound up with the mind's capacity for self-repair and the preservation of ego function.

Conclusion

In conclusion, *With Climate in Mind* reveals the extent to which we are primed to relate – both consciously and unconsciously – to the familial (more intimate) climate, the physical climate, and the (wider) social climate. Each is different and interacts with and influences the others, and all are essential to our safety and survival. In this sense, they are primary objects of both attachment and scrutiny.

Climate depends on the macro-level forces that shape current conditions, which makes it challenging to understand – not least because we so easily misread external reality through splitting and disavowal. Understanding a particular climate takes time and requires a framework grounded in care and the pursuit of truth. We do not experience a climate directly; rather, we feel its effects as what we might call "weather", which we are primed to read closely in order to survive and, hopefully, to thrive. We read the familial weather (e.g. "Is Mum or my analyst containing or frazzled today?"), the physical weather (e.g. "Am I hot or cold? Is the air breathable?"), and the social weather (e.g. "Is she showing signs of being caught up in social prejudices that lead her to elevate or denigrate me?"). We are primed to scan different types of weather and to try to understand the climates that produce it. In this way, we search for meaning.

Broadening and deepening our understanding of the climate changes the way we see the mind by illuminating the sheer complexity of its functions and tasks. This, in turn, can pose fresh challenges for the clinician trying to make sense of the material in a session. While it has always been difficult to understand such material, it may now seem even harder. The authors of the chapters in this collection show us how they navigate the climate in all its complexity, while still observing Freud's key principles governing the life of the mind: that it is the site of a ceaseless struggle between life and death, external reality and wishful fantasy, and that much of mental life remains unconscious. These principles still stand like lighthouses, guiding our search to understand the mind.

Notes

1 See Rosemary Randall, "Great Expectations: The Psychodynamics of Ecological Debt", in *Engaging with Climate Change*, ed. by Sally Weintrobe (Routledge, 2012), pp. 87–102.
2 Sally Weintrobe, *Psychological Roots of the Climate Crisis: Neoliberal Exceptionalism and the Culture of Uncare* (Bloomsbury, 2021).
3 Joseph Sandler, "The Background of Safety", *International Journal of Psychoanalysis*, 41 (1960), pp. 352–56.
4 Donna Haraway, *Staying with the Trouble: Making Kin in the Chthulucene* (Duke University Press, 2016).
5 Václav Havel, *Living in Truth: Twenty-Two Essays Published on the Occasion of the Award of the Erasmus Prize to Václav Havel*, ed. by Jan Vladislav (Faber and Faber, 1987).
6 Sally Weintrobe, *Psychological Roots of the Climate Crisis: Neoliberal Exceptionalism and the Culture of Uncare* (Bloomsbury, 2021).

Part I
Mainly Clinical

1 Keeping the Ecological Catastrophe in Mind

Delaram Habibi-Kohlen

The World We Live In: Cynicism, Exceptionalism, and Greed

We live in a world that makes us increasingly insecure. Shaken by polycrises – the turn to the right with attacks on human rights; a global pandemic and the rise of conspiracy theories; the wars in Ukraine, Israel, and Gaza; and of course the ecological catastrophe – nothing seems safe anymore. These crises appear to intensify and accelerate, gaining momentum as they multiply.

Today's generations, growing up in post-modernity within industrialized countries – where mass serial production and digitalization were taken for granted, capitalism accelerated, and many people, especially after the collapse of the Soviet Union and the Cold War, believed in a prosperous and secure future – have been shaped by these processes, which have taken root in our psychological structures.

Waiting, for example, has become extremely difficult in the age of internet commerce, and there is an unwritten law of instant availability due to mobile phones and email. Our subordination to the ideal of permanent performance is criticized by B. C. Han, who emphasizes how a person engaged in multitasking or playing computer games cannot indulge in contemplation. Rather, this activity "produces a broad, but flat attention which resembles the vigilance of a wild animal" – always ready to escape or run away.[1] Han calls for more boredom as a prerequisite for any creative process. In postmodernity, however, most people have no time for boredom. There is no time to immerse oneself in the natural environment or to experience oneself as part of it. For most people in industrialized countries, nature is disconnected from the experience of the changing seasons and cycles of which we are a part, or from our dependence on food and water. The division of labour and industrialization have taken us far away from this. Nature is now something simply to be cultivated: forests for timber production with footpaths; fields and meadows for agriculture; parks and gardens for recreation; mountains for skiing and hiking; oceans for shipping; skies for air routes. In short, human beings put nature to their own use.

DOI: 10.4324/9781003634959-3

Alienation from nature affects our hopes for the future. It uproots us. The polycrisis – and climate change in particular – generates uncertainty about the future. However, fears and guilt are often unconscious and defended against. One way of responding to these fears, it seems, is to become cynical or to content ourselves with "business as usual" – a phrase that Paul Hoggett describes as developing out of

> instrumental rationality [which] splits means from ends, privileging the "how" of things whilst ignoring the "why". It generates a world without meaning or worth. It follows that nature and humanity only have extrinsic value, they are "means" to an end (to generate profit and growth), not meaningful in themselves. And for this reason, capitalism itself veers constantly towards nihilism for [. . .] it knows the price of everything and the values of nothing.[2]

As Hoggett states, in postmodernity, nihilism takes the form of cynicism. Cynicism arises when it is too painful to hold on to our belief in the possibility of a better world. Its most destructive form is malignant cynicism which

> views all human virtues with suspicion, sees hypocrisy everywhere and has nothing but derision for liberal bleeding hearts. Because the person who tries to live their life according to moral ideals will inevitably fall short (moral failure is intrinsic to being human), the true cynic holds that anyone who is not amoral like himself is necessarily a hypocrite. The cynic takes his own state of de-moralisation as a strength. Ideals are for the weak minded. [. . .] The cynic has no faith, particularly no faith in human goodness [. . .] and he says this is "realism".[3]

This is a kind of cynicism that has changed from its original form. Modern cynicism does not remain aloof – it devalues empathy and has gained momentum with the rise of advanced capitalism, starting with Margaret Thatcher's declaration that there is no such thing as society, and her espousal of individualism and the idea that everyone is responsible only for themselves and their immediate families. The collective superego that might once have urged us to look out for one another is now out of fashion. Psychoanalysts, under the influence of our culture, also seem to consider the superego less important. A search on PEP web, the search engine of the Psychoanalytic Publishing Association, for example, reveals that most of the contributions under the keyword "superego" are from 40 years ago.

Perhaps the culturally deformed superego is more ego-destructive, envious, and murderous towards the ego because today's ideals are also more anti-libidinal, narcissistic, or even psychotic. Sally Weintrobe describes exceptionalism as part of an inner constitution shaped by a corrupted superego and a pathological organization that has infiltrated our culture and has been made manifest through the economic, political, and technical opportunities we now have for self-expansion and for acting out our greed.[4] Weintrobe emphasizes the destructiveness of greed when it becomes one-sided and relates only to the destruction of boundaries, rather than to the search for reality and truth. One might argue, for example, that it is vital for social justice, social balance, and peace to invest money in education, health care, increased taxes on the rich, or green technology rather than in the fossil fuel industry. But the warnings of scientists are devalued, ridiculed, and ignored. Weintrobe stresses that it is not greed alone that is destructive. Rather, the stability of the pathological organization is generated by the self-idealization and the sense of entitlement that greed, avarice, and deception potentiate.

It is interesting to see how the connotations of making money have changed over the decades – in the 1970s, it was not seen as being against the interests of the common good. Companies were prepared to pay taxes and offer more decent wages, and CEO pay was not as obscenely high as it is today. Greed was seen as an undesirable character trait, and seeking one's own advantage was not considered a virtue. When an exploitative practice – for example, within a company – is labelled as doing good for the inhabitants of a poor country, it shifts the baseline of what is seen as acceptable, and this assuages guilt. It is our culture that idealizes greed and collectively builds new pathological organizations that once again invade our personality structures. As the Australian sociologist Susan Long says: the deadly sins of the Middle Ages have become the virtues of today. Greed, for example, has been idealized to the extreme.[5]

External and Internal Reality in the Psychoanalytic Discourse

When thinking about clinical issues in relation to the climate crisis, one tends to encounter an internal division: the inner world of the patient on one side and, on the other, the external sphere with its political, social, and cultural dimensions. This division helps us understand our analysands in their innermost being, as well as the impact of the external world – whose exclusion is often necessary to maintain the ambiguity of the analytical situation. This is particularly important in times of global crisis. But there is a source of confusion in this: if the so-called external world is increasingly becoming a world of fantasy, filled with manic and omnipotent narratives spread on TikTok, Instagram, and other social media, how can one distinguish between fantasy – even psychotic fantasy – and reality? During the sessions they spend with us, most of our patients are preoccupied with their everyday lives. Recently, however, more and more colleagues have reported that patients are increasingly talking about their anxiety about the upcoming presidential elections, climate change, false facts, and a general fear of the collapse of the world. Does this require not only a different way of understanding the role of the outside world in our patients' lives, but also a new way of framing experience?

In *The Concept of Psychic Reality and Related Problems* (1985), Jacob Arlow rejected the concept of psychic reality as unwarranted – meaning that there is no such thing as a perception of the external world without an idiosyncratic transformation derived from unconscious fantasy. In this early study, Arlow argued that even the "here-and-now behaviour" of the analyst does not induce or alter the transference, but rather serves as a kind of daily residue of the unconscious fantasy.[6] Following him, it becomes clear that our task in psychoanalysis is to find this unconscious fantasy in our patients' productions and to reveal its individual meaning. This is further explored in *External Reality: The Elusive Dimension of Psychoanalysis* (1996), where Marion Michel Oliner emphasizes that patients do not need reassurance from the analyst about the "realness" of a situation in transference interpretation, since external reality functions as a rock of reassurance in itself – either patients can use it as a defence against unbearable guilt or they cannot. In her view, the distinction between internal and external reality should never be abandoned by analysts, but there is no need to "reintroduce external reality into their treatments". She argues that this demand to bring back external reality into the consulting room results from the conviction that psychoanalysis

> can and frequently does become the search for the culprit, be it in history, in external reality, within the patients, or, especially lately, within analysts themselves. But an assessment of external reality based on the need to find the culprit for the individual's difficulties leads to a very unbalanced picture of pathogenesis in which the patient is a victim and nothing but. It does not advance the study of external reality in its complex interplay with the mind.[7]

This chapter was written at a time when the role of external reality was being hotly debated between self psychologists and object relations theorists, as was the role of trauma vs. unconscious fantasy. Oliner emphasizes that her intention was to

> highlight the dynamics of the relief that is inherent in the trust in an orderly, impersonal world, in which one neither has control nor bears any responsibility; and I intend to stress how such a reality limits omnipotence and alleviates guilt. [. . .] I intend to show that unconscious guilt has an impact on all spheres of thought.[8]

She quotes Arlow's suggestion that there seems to be an underlying unconscious fantasy in both patients and psychoanalysts that psychoanalysis is about finding the true culprit of a crime, in which neurosis is the crime and the patient is the victim. Thus, the process of finding the cause becomes accusatory. Oliner claims that, under the pressure of "trauma theory" and the mainstream – as well as patients' accusations that the analyst is sadistic – an unrestricted analysis of unconscious fantasies is presented. One might say that this paradoxically reinforces unconscious guilt in both analyst and patient, since the aggressive and libidinal thrive, and the ambiguity of the analytic dyad seems dangerous and should be avoided.[9]

This relates to the issue of climate change and the implications of guilt: are psychoanalysts creating an unconscious enclave when they collude with the patient's blaming of external objects (the oil industry, financial markets, politicians, and so on) and turn away from the patient's own feelings of guilt? Similarly, if we look at the opposite scenario, are psychoanalysts participating in a mutual withdrawal when they search exclusively for the patient's unconscious guilt and fail to consider their own guilt associated with consumption and collusion with external objects? We must not forget that both the analyst and the patient live in a shared culture that promotes self-entitlement and emphasizes overconsumption as something that enhances self-esteem. Is there a way of seeing both the neurotic conflicts of the patient and the increasingly unstable – and, in many ways, false – external world, which creates new wishes and desires, selling them as real needs at the expense of the climate?

Harold Searles describes the relationship between humans and the natural world, with its dynamics of identification and projection. At an Oedipal level, he analyses the refusal to relinquish genital primacy and omnipotence when it comes to automobiles, as well as the refusal to obey a seemingly moralistic threat from an ecologist, who may represent an Oedipal rival.[10] Searles examines the fear, envy, and hatred towards these perceived rivals, focusing in particular on overconsumption and its impact on children and their futures. This is a depressive position, characterized by complex, paradoxical dynamics. He writes:

> We tend erroneously to assume that nothing can be done about the pollution of the present-day environment because of our deeper-lying despair at knowing that we cannot recapture the world of our childhood and at sensing, moreover, that we are retrospectively idealizing the deprived and otherwise painful aspects of it. The pollution serves to maintain an illusion in us that an unspoiled, ideal childhood is still there, still obtainable, could we but bestir ourselves and clear away what spoils and obscures its purity. In this sense, pollutants unconsciously represent remnants of the past to which we are clinging, transference-distortions which permeate our present environment, shielding us from feeling the poignancy of past losses, but by the same token barring us from living in full current reality.[11]

The argument is that humanity will continue to repeat the same ongoing destruction of nature in order to avoid guilt and mourning for a lost paradise that can never be restored. This is especially true as inseparable, interpersonal, and enduring relationships outside the home are vanishing. We are destroying the planet, Searles argues, to ensure that we have nothing to lose when we eventually die – this is, after all, the working of destructive envy. The modern world is so overwhelming that regression is the only way to cope. As Searles argues, "this 'outer' reality is psychologically as much a part of us as its poisonous waste products are part of our physical selves".[12] Technology and nature are perceived to be in direct connection with the individual's internal object world.

Luc Magnenat takes this further, stating that our destructiveness, infantile sexuality, and "non-thinking" are externalized and shape our external world.[13] This manic and omnipotent externalization is a defence against our fears of death and collapse, which invade and influence both our external and internal environments. In agreement with Searles, Magnenat sees the apathetic attitude towards a destroyed nature as an Oedipal rivalry – a profound disappointment in our parents, who left us in an overcrowded and depleted world with too many siblings. We repeat with future generations what has been done to us, and as a result, Magnenat sees people in today's industrialized world as narcissistic and armoured against pain. In this process of destruction, however, humans begin to identify with nature and fall prey to a kind of melancholia. When we neglect nature, we destroy a crucial aspect of ourselves – not only in terms of vital resources but also in terms of our emotional foundations. As a result of our own destructive tendencies, we treat nature as uncaring and indifferent – a thing to be used – thus denying our dependence on it and further severing the bonds with other living things, including each other. We are caught in a vicious circle: we deny our need for the natural world, continue to exploit it greedily, and deny our actions in an effort to manage our guilt and stave off grief.

Climate Change and Psychoanalytic Treatment

In criticizing the psychoanalytic community, Searles states:

> My hypothesis is that man is hampered in his meeting of this environmental crisis by a severe and pervasive apathy which is based largely upon feelings and attitudes of which he is unconscious. The lack of analytic literature about this subject suggests to me that we analysts are in the grip of this common apathy. But a second factor, a special felt hazard in our profession, tends to inhibit us from making the special contributions we could make: we fear that an active concern with this present subject will evoke, from our colleagues, nothing more than a diagnostic interest as to whether we are suffering from psychotic depression or paranoid schizophrenia.[14]

However, since Searles' work in the 1970s, things have changed. There is an increasing number of psychoanalytic papers on the subject of environmental destruction, and many activities within the psychoanalytic community aim to raise awareness and create space for debate – but it is still not easy to keep external reality and its impact on the psyche in mind when working with individuals. Even though many psychoanalysts claim to think about the historical context of each moment in their interactions with patients, it remains challenging to resist a perspective that views the mind as ahistorical, timeless, and anthropogenetically centred – as though we are dealing with the timeless structures of the psyche and the same key questions of existence across time. To believe that the problems of humanity have always been the same and will never change may give comfort and stability to some psychoanalysts, helping them feel safe in the belief that – however much the current way of life and its values may change – the core of human functioning will always remain the same. Psychoanalysts may fall prey to self-deception when we tell ourselves that we are dealing only with inner conflicts. We protect ourselves from anxiety when the subject of environmental destruction arises by treating it like any other intrapsychic conflict, falsely reassuring ourselves that if we "stay with the theory", we won't be too affected by the patient's fears and pain. In other words, we can use psychoanalytic theorizing as a defence against knowledge of external reality and its effects on us. If this is true, then I suspect we are filled with discomfort when we encounter patients who are concerned about the ecological crisis. What they bring with them confronts us with complex feelings of anxiety, guilt, fear, and often overwhelming helplessness. After all, we live in the same world, have developed along similar cultural lines, and many in our profession are children of affluence, born into an era of industrialization and commerce. The uncomfortable feelings we are confronted with may well be evoking our own insecurities.

By concentrating on the individual inner reality of the patient, we may transport ourselves to a place of apparent safety – away from the pain we experience when we realize we are in the same situation as the patient. They may be experiencing a terrible fear of helplessness and abandonment – a fundamental fear. If the patient experiences the analyst as accompanying them, talking to them about the fear of approaching forest fires or water shortages, they may not feel so lonely or trapped. The patient may feel they have a witness who confirms two things: that their fears are not pathological, and that they are not alone in them. However, if we dismiss the subject of climate catastrophe as unreal, as unfit for psychoanalysis, or if the patient notices us subtly changing the subject, they may feel abandoned – perhaps labelled as hysterical or as exaggerating. And if we are moved by fears of wildfires, droughts, storms, economic collapse, or mass migration, we may try to ward off our own anxiety by calling ourselves hysterical or exaggerated. In doing so, we are giving in to the urge to deny the reality of climate change.

Even though we may consider ourselves enlightened people – we, like so many others – are likely in denial about the reality of climate change. By distorting reality and breaking the link between actions and consequences, we avoid thinking about the climate emergency on our immediate horizon. We think about it in a more abstract way. It is happening somewhere else, and there is nothing we can do about it anyway. We may not think about mass migration, wars over water and land, food shortages caused by crop failures, or the increase in infectious diseases brought to us by insects that thrive in higher temperatures, happening on our own doorstep.

Projecting catastrophic fears into the future is another way of blurring the picture – it will happen then, not now. However, by going further, we can face this unpleasant reality and see it for what it is: humanity is involved, and we are unable to control the outcome of some of the physical processes we have unleashed. Knowing this can lead to withdrawal, cynicism, and depression. Facing the damage, however, can lead to a position of repair – if we can mourn the very fact of our limitations: that we cannot repair everything. This means mourning the loss of our omnipotence, which is especially difficult in the face of all the other losses we have to bear.

Trees: A Case Study

Patients in my sessions talk about trees as a way of connecting with nature, and in recent years, at least, trees have also become a symbol of the threat to nature. Trees symbolize security, and their physical structure is associated with feelings of solidity, strength, and continuity. They outlive us, comfort us, protect us from the heat, give us adventure when we climb them, and provide us with sustenance in many ways. We feel pain when trees are set on fire, or cut down, torn apart. For the first time in our lives, however, all of this seems to be in question.

My patient, Mr A, spoke of his fright at the sight of dead trees. Having been in analytic treatment for many years, for the last few months he had been seeing me once a week and even this was threatened by the financial difficulties he was facing after an accident. We talked a lot about his fear of the end of his analysis, and of a regressive pull into withdrawal, which had isolated him from his family but which often seemed to be his only way of coping with them. Mr A expressed how he felt less persecuted by hostile objects in the forest, which for him represents a retreat from a disordered world where he can find some freedom from an invasive and "rampaging" object. As Mr A recalls:

> I was in the forest with my children and we saw these hundreds of dead trees everywhere, everything all brown and bare. I suddenly had a shock at the thought that one day there might not be any trees left. I thought, maybe my children won't be able to experience

what I still knew and what always gave me comfort. At least in nature, I felt free. When my mother had yelled and screamed at me again and I could go out afterwards, I felt comforted. It was somehow that I felt connected and as part of an entity that carried me. If I couldn't stand people, nature was always a consolation.

I was struck by the shock that the patient described and by the world that he imagined: a world without trees is a world without protection, without security, without generational support. It is a world where nothing can grow, where life has stopped. This is the end of the transmission of experience from one generation to the next, which is reminiscent of the ending of his analysis and the feeling of being "cut off" from his analyst, returning to a world of dead and dying, truncated objects, and the terrible loss of his good object, signified by the trees.

The world he imagines – without the protection of shade and the possibility of shelter – is indeed bleak. I had been in the forest the day before and, like him, had seen hundreds of dead, brown trees. I had already lost my way because of the deforestation that had changed the familiar landscape beyond recognition. I too had been gripped by horror, flooded with the fear of what would happen if this continued. Will there be forest fires here, like those in Brandenburg, California, South America, Indonesia, or Canada? After a while, I say: "The forest and analysis, these are the guarantors of freedom and stability, both. I think you are afraid that both could be destructible and that then there is no future". To which Mr A replies: "With all the fear I always had. But I always had the feeling that some things are always there and always stay there, like the forest. It's terrible when the certainty goes away." This resonates with me. It is terrible when certainty disappears, and I do not feel particularly supportive, as I am also unsure if I can have hope or if hope is naive. Mr A and I are temporarily in the same depressive state. It explains his reaction to the trees, as well as his stability and sense of hope. We both know that he has never had much certainty about anything.

His frequent experiences of rejection by others are a reminder to me of the patient's frequently experienced guilt over aggression: as a child he had hit another child and was severely punished. As the son of an ill-tempered and violent father, his own contempt and suppressed anger toward his father are redirected towards himself. He identifies with a masochistic mother who has also caused him to feel guilty because he was never able to help her with her burdens, overwork, and suffering. In my countertransference, I often experienced the same guilt of not being useful enough to Mr A, along with my frustration at my patient for not allowing me to be useful. I often felt that this was a repetition of what he had experienced with his mother.

He had described a suicidal temptation to get it all over with. I say, "When the certainty goes away, would you sometimes prefer it all to be gone already, without this uncertainty?" To which Mr A replies, "Yes, at least then it wouldn't be disappointing."

On the surface, this could look like a normal session in which I interpret the patient's images of a walk in the forest as an expression of his fear of the end of his analysis. The disappointment then has to do with an angry object that haunts him and with which Mr A is both identified and afraid. I ask him:

> You fear losing the bonds that give you security and therefore you are the first to cut them down in order not to be disappointed. I wonder if planting new trees, little ones that need time to grow – would be too painful?

Mr A thinks about this until the end of the session. In the following weeks, he talks about his reluctance to let anyone in and finally comes to his bottled up aggression, which he did not want to feel in order to "not to be like my father". His reproach against the "tree killers" (the rich, the mindless, the corporations) has another dimension – it is also an accusation against his raging and inaccessible mother figure. As we have been working along these lines for many years, it is possible for him, in the long run, to accept the meaning of our work as mostly useful.

The conversation with Mr A about the forest became one of several turning points in his analysis. While it was a small but nuanced difference from the usual work, to me this nuance seems to be of crucial importance. This is because I am also engulfed by the horror that Mr A describes of the destruction of the nature that offered him a place of refuge. I feel that we are in the same situation, even though we are two different people with different biographies. Seeking this common ground makes me bear witness to his depression, which I share and empathize with in the face of the destroyed trees. These trees mean something different to each of us, but they are also something that gives us access to a common, albeit separate, experience. Trees suggest something protective, sustaining, older, always present – a guarantee of continuity and a connection to something bigger than ourselves. The danger that trees may disappear is a new one in our experience, one that both frightens us and that threatens what we have taken for granted.

Spiritual and material sustenance from nature and trees can often be a refuge from an intrusive object. One might also add that democracy, debate, and tolerance are now marked as transient. I maintain that I would not have been able to pick up this atmosphere and ask about planting new trees if I had not also felt the terror of the possibility of existential loss. This process went along with my inner confrontation with hopelessness and my search for resonance. This then helped me to talk about Mr A's aggression, as well as his nurturing and caring aspects.

In the context of this case study, and with reference to the earlier views of external and psychic reality, it can be argued that this patient cannot rely on turning to the external world as a defence. His psychic reality and the burnt, barren, dead landscapes of external reality merge indistinguishably. By naming the patient's need to destroy connections as a way of averting disappointment, and by asking him about new plants that need time to grow, I believe I am – in retrospect – introducing some new, distinctive lines into my patient's inner, devastated landscape. I think I was able to do this and bear the image of the dead trees because I had experienced the real horror of the outside landscape, and could therefore relate to the horror of the patient's psyche on a different scale. For me, the dead trees were not just a few dead trees, but rather the horror of a dead future. On this basis I was able to connect with my patient in a different way. I was able to feel a deeper pain, to feel more of his repressed aggression, and to follow his manoeuvrers not to go deeper into those feelings. Applying Searles' ideas to my patient, one could say that he avoids guilt (feeling the aggression towards a violent and destructive father), and he avoids grief by depressively clinging to a lost and idealized childhood but continuing the destruction because he is unable to grieve. He obeys the rule that tells him to stay in a passive mode, to isolate himself.

Conclusion

Humanity is capable of destroying itself not only by nuclear accidents and war, but also knowingly – and at the same time in denial – by exacerbating the ecological crisis through CO_2 emissions, waste, and the accelerated loss of biodiversity. This destruction is happening faster than we ever thought possible – we see millions of people dying by the end of this century, when the global South will be uninhabitable; we see social conflicts and the decline of democracy worldwide; and we see mass migration already underway. Acknowledging all this – and not dismissing it – is a shock to our systems. This kind of destruction on such a scale has never been possible before in history. What makes the shock even harder to bear is our own complicity in the damage done. Climate destruction and climate injustice are affecting everyone in the industrialized world. They are causing both conscious and unconscious feelings of guilt, which tend to raise the usual defences.

However, it must be recognized that psychoanalysts have moved away from the so-called external reality as being relevant to individual psychoanalytic treatment. The leap to the symbolic is sometimes a defensive one on the part of the psychoanalyst, working on the principle that if it is psychological, it is not so serious. If, for example, a patient's statement about refugees drowning in the Mediterranean is interpreted as his fear of not being carried by the maternal element or of being flooded either by maternal devouring impulses or by his own aggressive ones, then this may be true in terms of content, but it leaves out something crucial: the simultaneous horror at the extent of the real indifference in the world that is felt by both the patient and the analyst, and the hatred and guilt at having indirectly contributed to it. In this way, the internal and the external are distinguishable – otherwise we would all be psychotic. But they are also always intimately connected. One cannot be pitted against the other; they are always interrelated.

Psychoanalysis can enable patients to regain a buried relationship with the self and the world, so that symptoms are reduced, the view of reality becomes more bearable, self-esteem improves, and the capacity to care increases. It can also lead to a political self-understanding in the broadest sense. This develops through self-reflection and the capacity for disillusionment, reality, and self-knowledge that has grown in analysis. This can only happen if everything can be said without the patient experiencing rejection as a person (something patients are overtrained to detect in everything). However, the development of the person as a political subject in the narrower sense need not result from this. In the worst case, we have to accept that, for some patients, stagnation is the only tolerable state. This may be more secure for the patient's identity and life than change. The hope would then be that the patient's awareness of this regression might be the beginning of a slow process of change.

In any event, it is important for therapists of all schools to deal with their patients' statements about the world – the "outside" – and to talk with colleagues about how they understand and deal with these statements. The grim times in which we find ourselves, with regard to the "forces of nature", along with the violence between people and between people and nature, make this discussion more vital than ever.

Psychoanalysis has always had a cultural application, providing a lens through which we look at society and culture. This "beyond the consulting room" application is urgently needed. Psychoanalysis, with its knowledge of the laws of the unconscious, can explore how the structures of society change the structures of the individual, and how individuals in turn impact society through their behaviour. It can also explore how social tipping points occur. Awareness of the emergence of climate-damaging social structures needs to be promoted – for example, by including this topic in the curricula of future and practising therapists and analysts, as well as in inter-school discourse. We need to talk to each other – to avoid the consolidation of enemy images (such as "the political" vs "the apolitical", with each side negatively connoting the other). We also need to avoid a development in which psychoanalysts feel more and more helpless in the face of increasing real and traumatizing events in the world and on our own doorstep.

So-called external reality may bring some psychoanalysts into conflict with some of their habitual ways of thinking when they realize the shock of the environmental crisis – its speed and its consequences. If psychoanalysts compartmentalize their knowledge of climate change so they do not have it in the back of their minds to bring it up in sessions, we may end up individualizing our patients' problems rather than seeing the climate as something collectively overwhelming.

Then the analyst's multiple fears – of the natural world, of death, of disaster – can be deflected and disidentified. In Jonathan Lear's words, we need some radical hope which can look into the eye of future devastation without giving in completely to depression, and without succumbing to a manic belief that refuses to acknowledge reality.[15] This is rather ambitious, but in my view it is the only way to keep going.

Notes

1 B. C. Han, *Müdigkeitsgesellschaft* (Matthes und Seitz, 2013), p. 27.
2 Paul Hoggett, "Climate Change: From Denialism to Nihilism", in *Climate Psychology: A Matter of Life and Death*, ed. by Wendy Hollway, Paul Hoggett, Chris Robertson, and Sally Weintrobe (Phoenix, 2022), p. 29f.
3 Hoggett, "Climate Change: From Denialism to Nihilism", p. 31.
4 Sally Weintrobe, *Psychological Roots of the Climate Crisis* (Bloomsbury, 2021).
5 Susan Long, *The Perverse Organisation and Its Deadly Sins* (Routledge, 2019).
6 Jacob Arlow, "The Concept of Psychic Reality and Related Problems", *Journal of the American Psychoanalytic Association*, 33.3 (1985), pp. 521–35 (p. 531).
7 Marion Michel Oliner, "External Reality: The Elusive Dimension of Psychoanalysis", *Psychoanalytic Quarterly*, 65 (1996), pp. 267–300 (p. 275).
8 Oliner, p. 267.
9 See also M. Baranger and W. Baranger, "The Analytic Situation as a Dynamic Field", *International Journal of Psychoanalysis*, 89.4 (2008), pp. 795–826.
10 Harold Searles, "Unconscious Processes in Relation to the Environmental Crisis", *Psychoanalytic Review*, 59.3 (1972), pp. 361–74 (p. 363ff). See also Chapter 7 in this volume.
11 Searles, p. 365.
12 Searles, p. 367.
13 Luc Magnenat, "L'environnement non humain de Searles, revisité à la lumière de la théorie de la pensée de Bion et à l'ombre de la crise environnementale", *Revue Belge Psychanalyse*, 69.2 (2016), pp. 113–28.
14 Searles, "Unconscious Processes in Relation to the Environmental Crisis", p. 360f.
15 Jonathan Lear, *Radical Hope: Ethics in the Face of Cultural Devastation* (Harvard University Press, 2008).

2 Coming Alive in Relation to the Natural World

A Clinical Account

Lynne Zeavin

A little boy spent the first three years of his life in the American Southwest, a land marked by wide-open desert, with mountains the colour of red clay. The sky was vast, as was the silence. When he was three his parents and siblings moved across the country to Atlanta, Georgia – an urban place filled with people and the noise of traffic, streetlights, and garbage trucks – where the sky was hard to see. One day he asked his mother, "Mommy, are we still on earth?"

His companions – the red clay of the hillsides he climbed, the chokecherry and juniper trees, the brilliant sky, and the stars at night – were all missing. He did not recognize the urban landscape as earth, nor as home, for that matter. His early experience was a rather seamless integration of play, bodily sensation and movement, and connection and trust-building as he navigated the sensuous, evocative, and living surroundings of New Mexico. Now, in what felt like a remote and cut-off place, he longed for the companions who had been there from the start – as they always are, at least for those who have access to the outside world – the trees, dirt to dig in, the sky and stars, animals (stuffed or represented in picture books), to the water, sun, and moon. I think the story of this little boy shines a light on how, from very early in our lives, the natural world forms a part of each of us. Our early relationship with the natural world is part of what makes up our very beings – our bodily ego. Features of the natural world are among our earliest objects, and they help to forge links within us and between us. Think, for a moment, of the moon. The moon in one time zone is the same as the moon in another, only rising at different times. Little children routinely ask about the moon. My little grandson in California knows that "his" moon is the same moon I am seeing in New York. The moon provides a link.

***

The atmosphere of the pandemic pervades the clinical work I will describe. At the time this material emerged, COVID-19 was directly affecting us all, and its reverberations are still very much felt. In psychoanalysis, it has recast our frame. Many now work at a distance – on Zoom, on the telephone.

DOI: 10.4324/9781003634959-4

During the pandemic, disembodied as we were, we had to find new ways of making contact with our patients. We listened with ever greater care to the voice, to its latent expressions. The new technologies brought their own anxieties – ostensibly about the technology but often more deeply a reflection of the question of connection itself. Over this long four-year period, I have noticed the resurgence of very early anxieties, the particular narratives of which have formed the basis of our analytic work. The natural world has played a significant role in helping our patients – and ourselves – manage unmanageable anxieties about time and timelessness, about that which cannot be known, and ultimately about survival and death.

The case I will describe is of a woman (Ms S, now in her late fifties) whose internal affiliations have habitually been on the side of death. Ms S has been in analysis for over a decade. She is an executive who runs a large public works corporation. The title "Director" holds great importance for her. Over the years, we have explored her up/down view of the world, which, in the transference, has led to my "help" being experienced either as a wish to bring her down or as something that evokes an impulse to destructively negate or evade what I say to her. Certainly, her insistence on being "up" and untouchable for many years concealed and protected against a fear of contact with both her psychic reality and the outside world. I came to see that she was panicked by the idea of being apprehended by me. The patient lived – and still often lives – in a daydream world of violence, beating fantasies, and paranoid anxieties, where she anticipates and staves off attack. Ms S grew up by the sea. She talks of loving to swim, of feeling at home in the ocean, even when it is wild. Others have referred to her as a water skater, as – like the insect – she is always restless and buzzing, always on the move.

What I want to consider is how repair and the capacity to make reparation are crucial to psychic survival and intricately linked with the development of a capacity for concern. I want to describe how my patient came to relate to the natural world without the sense of danger and threat that accompanied her involvements with people, and how her relationships with the natural world offered an arena in which she could feel "more human". Towards people, she felt suspicion; she was always on her guard, wearing a broad pre-emptive smile that concealed her fearful and vigilant stance. But her early relationship with nature – first with the ocean, then gardening and a cat – began to help her move out of her protective, omnipotent shell and ease some of her wariness. Her engagement with the non-human world has ushered in a greater capacity to tolerate depressive feelings and a move toward reparation.

Edna O'Shaughnessy has described reparation in this way:

> Reparation is an impulse of Eros. In Freud's words [. . .], Eros is "the instinct to preserve living substance and to join it into ever larger

units". In the depressive position, the ego attempts to do just this with its objects and itself – to preserve or restore life and to integrate the different aspects of self and object. [. . .] As the ego becomes more integrated, what happens to the discontents of civilization that Freud described – the renunciations of instinctual gratification and the inevitability of guilt? Instinctual renunciation could become a loss to be mourned and endured, less from fear, more from empathy with others. And, if things go well, a humanist conscience will develop.

Whereas Freud saw the discontents of civilization as arising from a restriction of the instincts, Klein viewed unhappiness as coming from the ego's identifications with its objects. O'Shaughnessy goes on:

> if these are felt to have been humiliated, damaged, even murdered, the ego feels guilt and pain and in identification is mortified, broken, and deadened. If the guilt and pain of hating primary objects can be confronted and to an extent worked through, and some reparation made, the depression of containing damaged objects changes to a sense that there are living objects within.[1]

For Ms S, the difficulties of coming alive were enormous. For many years, the natural world had offered a haven and a place of safety. Although she had, for years, made her way by preserving her fidelity to a dead and depressed internal maternal figure, the natural world provided a space where she could make investments rooted in a feeling of care and a pleasure in communion. Coming alive means a greater tolerance of separateness, and for my patient, it explicitly meant being able to separate from the deadness of her internal object; it meant tolerating being small and becoming able to receive. It also has meant becoming receptive to psychoanalysis, which turned out to be no small feat. Psychoanalysis, after all, entails submitting to help and facing one's own limits – and my patient found this humiliating. She came to me moored in a closed world of dead and deadening objects. She identified with these damaged objects and had a strong erotic attachment to her own destructive fantasies. She retreated – regularly – into a punishing world, a world where, as she often put it, she had to "crack the whip" lest she fall down on the job. In the course of this long treatment, by addressing Ms S's destructive identifications – particularly in the transference – we have been able to sense a diminishment of her fear and dread, and, gradually, the development of a feeling of concern for the non-human figures in her world. Her affiliation with the natural world, rooted in an early relationship with the ocean, has allowed her to experience care rather than fear, and to develop greater trust in both herself and elements of her world. Concern for her plants and an animal has fostered a calmer internal state from which she feels more generative and less damaging. With a diminished internal sense of damage, she finds herself more able to bear her own guilt and suffering, which has paved the way for a more genuine involvement with non-human beings and, along with this, a move towards reparation.

The reparative urge is meant to resuscitate the damaged or dying internal object until it can be integrated and claimed as one's own good object. For Melanie Klein, the good object is the outcome of a long struggle between the destructive and loving forces within the personality, and it is meant to represent a source of love, Eros, and stability. Ms S had been without a good object for a very long time. There are no reports of a grandmother or anyone else coming to her aid. She had an isolated and lonely child-hood. Although there was great financial wealth, it was earned and lost many times over, producing a manic world followed by wreckage. Living in a huge house (55 rooms, she told me) overlooking the ocean, she felt a huge gulf between herself and her older sister – a much-admired girl who provoked envy, rivalrous longing, and resentment. Only when a younger brother was born, 10 years after my patient, did she feel she could experi-ence love.

A little bit of history here is important to understand more about Ms S's own life-and-death struggle, and the role the natural world has played in helping her turn towards repair and life. She was the third child born to her parents – but the baby who came just before her had died of a congenital abnormality at four months old. Her mother was advised to get pregnant again as soon as possible. With a lifelong feeling of being unwelcome in the world, my patient has always felt her primary objects to be either tormentingly absent or critical and reproachful when present. A photograph she often refers to shows her, aged 3 or 4, staring up at her mother, who does not seem to notice her presence. Her way of dealing with early anxiety and dread about being and non-being has been to try to position herself *above*. The exception to this, of course, was her time spent outside – playing, exploring, often alone, and often swimming in the sea – this being a proto-good object that has through her analysis sta-bilized more securely.

Ignês Sodré describes the state of mind in which unbearable persecu-tory guilt attacks a person's sense of their own goodness.[2] For Sodré, guilt becomes unbearable when reparation is impossible. Reparation is felt to be impossible when the object is too damaged, too depressed, or too vengeful. After many years in analysis, in what I see as a movement towards repair, Ms S has begun to engage more reliably with the natural world – a world that, as I have described, was deeply embedded in her early internal land-scape. Ms S has deepened her love of her garden and, most importantly, after years of discussion, she adopted a cat.

Clinical Material

One Friday, Ms S concluded a painful session by telling me that a songbird had appeared at her window. She found this surprising and beautiful. She said, "Usually there are so few birds this time of year who sing." I was struck by this, and I said that I thought she felt comforted by the presence of this bird – that it felt like a spot of life she could sense and allow in, momentarily reassuring. She said that yes, the presence of the bird had made some of the difficult feelings less powerful.

Like with the songbird, she seemed able, just then, to receive something from me without taking over my function or using it to fuel her own self-sufficiency. It also allowed her to avoid the familiar cycle of terrible excitement, in which she moved rapidly between victim and perpetrator. She was able to appreciate contact, much as she had appreciated the songbird who had visited.

That night, she had a dream:

> I am walking with B (younger brother) through a forest. We see that it is a cat refuge. Or it is supposed to be. But I see a man flailing [*sic*] a cat with a rope, like a whip, he is beating this cat, saying "get down" and the cat is going higher and higher in the tree. It is violent. But then I see another cat, and I say psss psss psss and the cat runs to me, I was looking at her, it was Beatrice (her cat in real life). She recognized me, she came to me, she came back to me, she recognized me and I was so grateful that she recognized me.
>
> She ran. She couldn't run for a long time, not when she got sick, and not so much around our house, there wasn't space. It was such a comfort. I lost her, and then I regained her. I woke up and I thought she really is gone. I have lost her, I miss her so much.

She went on:

> So the dream isn't true. She is not alive. But beating is true. Abuse is true. It was a supposed sanctuary, but that wasn't true. We are working on a colloquium about bugs and beasts. Our relationships with non-human beings. And their possibilities of being. I feel disconnected from my own possibility of being and somehow Beatrice represents some sense of goodness that I can trust.

I said I thought she had found some relief from being very high up and from the firm ground of beating, and I added, "I think you are aware of strong feelings of sadness and loss both when you woke up – and also yesterday, as you were describing how hurt and vulnerable you felt." She said, "Yes, those *are* true feelings of being – hurting, afraid of hurting. Loss. But I don't really know how to have a conversation or contact that is not a beating." To this I replied:

> The sanctuary is a "supposed" sanctuary that you withdraw into – where beating goes on – but in this dream, Beatrice runs towards

you, does make contact with you, and you feel recognized. I think her recognition feels safe and acceptable, and you're able to connect.

She responded: "It gave me incredible comfort to have that contact. Something was restored. My humanity, I think. Something felt alive in a different way. Funny, because I do know she is dead." Her thoughts then turned to the session the day before: "Yesterday I also felt recognized by you; I felt soft and not in that bad way, not in the weak way, but soft like I could let something in. I didn't scramble, didn't move to organize, compartmentalize, I took it in."

The dream conveys an important aspect of Ms S's predicament. She has been living in a shelter – a supposed sanctuary – where she is in retreat from a more engaged contact and from the paranoid and depressive anguish that such contact brings. She lives and relives unbearable guilt over early violence (her father's beating of her sister, and her own triumphant fantasies that are played out again and again), along with her painful wishes to be let in and to belong. She hates her own yearnings and always strives to overcome them. Violence, which brings guilt, also brings relief from unbearable longing. Ms S has sheltered her mind in an omnipotent and persecutory world that protects her from having to contend with her own violent feelings, along with the knowledge of her need for, and dependence on, her objects. The world of objects that she needs is also the world that she reviles and fears. The two parts of the dream visually represent a split, and the fact that the split has become so representable is, of course, hopeful: she has lived for such a long time in a sadomasochistic refuge. She has an unconscious sense that such a refuge protects her from living in a more humane world, with all its disappointments, threats, and losses. She needs her refuge because it protects her – and others – from her feelings of fury and violence, as well as from her tremendous need. When she feels less threatened, she can emerge briefly from the refuge and have a more open connection with her analyst, which may include – even if only briefly – a sense of a more reliable internal and external world. It is meaningful that what allows her to feel recognized in the dream is a cat – her cat, Beatrice – whom she had loved and cared for in real life, and with whom she had been genuinely vulnerable for many years.

As Ms S talks about her softer response and taking something in, she notices a reduction in her manic defences – which may allow a diminishment of paranoia and guilt. She notices that she does not have to scramble, which is her usual sense of having to keep moving, like a water skater. Sodré describes a particular form of manic defence not only against terrible persecutory guilt but also against the rise of depressive feeling. Manic defences arise in part to block out depressive anxieties and feelings of need-as-persecution. They restore a more omnipotent position. In her dream, when she sees Beatrice, she is able to experience attachment and pleasure. There is grief upon waking. There is a feeling of longing which she can sustain while also facing loss, which then enables her to have more open contact with a real and living object. This allows her, momentarily, to take in something from her analyst that she experiences as good, without having to destroy it.

A few days later, towards the end of a rather disorganized Monday session, Ms S exclaimed:

> Oh My God, I forgot until just this moment that, over the weekend, we discovered that one of the neighbour's gardeners – it's a new neighbour – butchered one of my precious shrubs – it was beautiful – I have been tending and caring for it – shaping it since it was small, maybe for six or eight years.
> Someone actually destroyed it.
> I am so upset.
> I saw a huge yard waste bag filled with the branches.
> I think their gardener must have "taken it down".

When I first met Ms S – and often since – she described a pleasure taken in the natural world: plants, the ocean, and gardening. Equally, she has often mentioned her conviction that people want "to take her down". Here, her shrub has quite literally been taken down and is now ruined. Her attitude towards this revealed a position she has long assumed. She described how, after the event, she and her husband went to the neighbour's house: "we were masked". She introduced herself and, in a very polite, restrained way, told them what had happened. They felt terrible and were instantly apologetic. She had not disclosed how furious she was. In the day or so before this Monday hour, she had become more and more disorganized, spiralling down into her familiar well of blame and self-flagellation. After describing this state, she recalled the shrub, and with this we were able to make more sense of her pull into a familiar agitated state where she is absorbed in conflict which helps her avoid an experience of helplessness and loss. Her usual sanctuary is now a supposed sanctuary; she knows that it does not actually protect her. As we talked about the link between the loss of her shrub, her hatred of the neighbours, and the fury she felt she must restrain, Ms S could see that she was turning all of this against herself. In that frame of mind, she could neither mourn the loss of the shrub nor imagine repair. She could not even let them see she was upset, let alone allow them to offer her a solution. By the end of this hour, she came to say, "It is not dead, it is butchered and mangled, but not dead" – a clear representation also of her internal world. She then thought of her friend E, who is a "master gardener", whose help she might be able to enlist. Maybe E would help her figure out how to revive her shrub and aid in its healing. She thought she would be able to ask her for help. We can wonder about this – will it be a form of manic repair (E is a "master gardener", after all), or might Ms S be able to receive help and face the loss she has felt? This makes all the difference.

The following day, Ms S recounted her other losses, including her own desire not to have children. She had decided against having a child out of fear of her own destructiveness. She could not imagine herself capable of giving birth to something living and, if she could, she was sure she would destroy it. It is precisely this history that made being able to adopt and raise a cat so significant. Ms S was able to care lovingly for her cat; she would speak to the cat, and she tended to it during its final months of life. In the dream, she is able to revisit the cat and have a different access to grieving.

Lindsay Clarkson has written:

We are not living in good contact with reality when we act as though we are self-sufficient. Sanity founders if the sustaining aspects of caring relationships between and among humans are treated with contempt, or if one is too paranoid to make any use of them.[3]

Certainly, for many years, this has been Ms S's psychic predicament.

Living in a deadened internal world, fearful of her own needs and vulnerability, as well as her hateful feelings and violent urges, she has retreated to a high-up place which, interestingly, has made her feel she has no place among humans – she must be superhuman. Very recently, alongside the tenderness she has expressed for her cat, she has begun to desire a place among humans, starting with her own extended family (particularly the grown adult children of her sister and their children). More reliably, the non-human world is one in which she can imagine attachments that are meaningful to her. She can care about nature and find comfort in a walk through the forest or a swim in the ocean; she tends to her garden and speaks of having another cat. This allows her to escape from the up/down world she has needed to inhabit for so long. She can feel the possibility of her own investment in another. These interests also seem to potentiate a capacity for stillness.

When she said that her dream restored her humanity, she meant that for a moment she was able to feel part of an ordinary world of living beings – not up or down, but within and part of it.

The world of nature is such a world, and it allows Ms S a domain of trust and care that is rarely available to her in the world of human interaction. For her, the restoration of her humanity means that she can take her place in a living world of plants and animals, and that she can express and feel, however intermittently, concern and interest in their care. This is how Ms S describes her own relationship with the external world:

I had come to feel that there is nothing loving in the world, no love I can trust. That's always been the case. I guess I haven't tested it. I felt my own survival depended on never trusting. Now I see it a bit differently. When J (a former boyfriend) and I broke up I felt like a desiccated plant. It was if there were no water, no Light. As if all the roses around the house were suddenly near dead, the irrigation system was broken. Now I have a garden, I can tend it. But I don't tend it alone, I do it with D (husband). It doesn't keep it from being vulnerable to the elements, of course, or neighbours. But I know I have a capacity to tend my garden and that things can grow and thrive. That must mean something.

And she added, crucially: "And I know, even though I resist it, you are here to help me." For Ms S, a burgeoning relationship with the natural world is precisely what has come alive in the course of her analysis – and these involvements provide an area where she can feel more of a sense of dependency. She knows that she tends the garden with her husband, that irrigation is necessary for life, that damage is part of it, and that reparation is possible. The natural world provides an intermediate zone where it is safer to rely on another, and where love and concern can begin to develop. These concerns for the wellbeing of her garden and for her cat's welfare can, for brief moments, allow for a different feeling towards her analyst; she can come out of her lifelong refuge. She notes that this is not without potential risk or vulnerability; it is not, at this moment, an omnipotent solution. She can acknowledge the place and value of others and turn towards a world of living companions – plants and animals – that both nurture her and allow her to feel more alive. In describing reparation, Henri Rey explicitly links it with nature: "Reparation is a process so widely active in nature. [. . .] We are familiar with the renewal of life when spring repairs the winter destruction and life reappears vigorously creating new life everywhere."[4] Many aspects of the natural world – trees, plants, animals – speak to Ms S and allow her to stay still for a while and to feel alive in her stillness. Lisa Baraitser, drawing on the work of Christina Sharpe, has inquired whether a capacity for stillness may in fact be understood as an important psychic achievement. She writes,

> Passivity understood as the mental capacity for withdrawing, for instance from rampant consumption and unsustainable growth, is vital if we are to interrupt the business as usual of what some term "heteropatriarchal racial capital" as it burns up the earth.[5]

A relationship with the natural world, when sustained, ushers in a different relationship with and valuation of life's offerings. Even if it does not yet translate for Ms S into ease with other people, her contact with the natural world provides a very personal form of engagement. She can tolerate her smallness without undue rivalry. The trees do not evoke feelings of shame or of envy. Instead, she can value the sense that life goes on, that the natural world endures, which not only relieves her of some of the anxiety about her own destructive wishes, but also allows a channel for the expression of reparative desires and a capacity to receive care.

Notes

1 Edna O'Shaughnessy, *On Gratitude in Envy and Gratitude Revisited*, ed. by P. Roth and A. Lemma (Routledge, 2008), pp. 79–91.
2 See Ignês Sodré, *Imaginary Existences: A Psychoanalytic Exploration of Phantasy, Fiction, Dreams and Daydreams*, ed. by P. Roth (Routledge, 2015).
3 Lindsay Clarkson, "Trees and Other Psychoanalytic Matters", in *Hating, Abhorring and Wishing to Destroy*, ed. by Don Moss and Lynne Zeavin (The New Library of Psychoanalysis, 2022), pp. 217–33 (p. 217).
4 Henri Rey, "That Which Patients Bring to Analysis", *International Journal of Psychoanalysis*, 69 (1988), pp. 457–470.
5 Lisa Baraitser, *Enduring Time* (Bloomsbury, 2023), p. 917.

3 The Colossal Divide
Transference-Countertransference Crossfire

Karyn Todes

Fault Lines: Where Heartbreaks Become Chasms

Two serious symptoms can arise out of soul murder or childhood abuse, namely depersonalization and/or derealization.[1,2] A child who is emotionally annihilated does not feel himself to be real. It is as if a glass plate separates him from what is going on around him, and he feels assaulted by strange experiences or a disturbed sense of reality violently at odds with his own. Imagine a fault in the earth's crust, slicing through the fractured landscape. Rodney stood at the edge of this fault, where the ground abruptly dropped away into a deep, jagged crevice. If the earth's crust were severed into two halves – forcefully torn apart by the relentless movements of tectonic plates – separate formations would emerge on either side of the fault line. Heartbreaks became chasms, creating these "fault lines" that divided his psychic world in two, and this point of division became the focus of his analysis. I call this "the colossal divide".

Perversion is a subject that covers a wide spectrum of ideas in psychoanalysis. I offer my own way of looking at it. To be alive means to be able to feel pain, but the perverse patient does whatever he can to ensure he does not re-establish a life based on relationships. He creates an altered, delusional reality that ensures he stays as far away as possible from external reality and people. He makes other people suffer his pain by fighting them, trying to ensure they experience his sense of death, "killing" them as living requires establishing authentic human connection. This cruelty ensures that he keeps the connection to the toxic, abusive object alive inside himself by inviting the object to return the cruelty. In other words, he becomes a murderer seeking revenge for the death imposed upon him. If he were to face the reality of his own inner death, it would mean addressing the original trauma from which he sought refuge when he "disappeared" into isolation under the weight of psychotic anxieties. He continues to wall off his lifeless self so that he might never again encounter the source of his pain. The extreme measures taken in the service of self-defence lead to a perverse outlook, derived from unconsciously creating a delusional reality in which he tries to involve

DOI: 10.4324/9781003634959-5

others – hoping to avoid total isolation while also avoiding confronting his internal horror.

Rodney lived in a world of action where feelings and thoughts had become facts. He did not know how to relate to anybody. He simply pinned his own perverse reality – a particular outcome of heartbreak – onto other people, interpreting his life according to his delusions. Soul murder is underpinned by profound loss and a failure to develop. In these circumstances of overwhelming pain, Rodney froze his suffering, shutting it away in a dark chasm.

Cracks of Chaos

Rodney's mother was admitted to a psychiatric hospital after his birth, and a series of babysitters replaced her. He described his father as "a drug addict who beat and punished me severely".

Rodney's father was like a deadly predator in the wild – hunting young Rodney, crushing and terrorizing him, punishing him, and eventually eliminating him as a viable human being. The scale of inequality between Rodney's helpless infant self and his cruel father was intolerable, which is why the analysis felt unbearable. I felt like I was a young Rodney, and Rodney, in turn, felt to me like a version of his father. He trusted nobody; he had no friends; life was as barren as the barest desert. In response to this treacherous childhood – and given that he had never known what life was about or had any sense of what people are really for – Rodney believed that all people were dangerous and to be avoided at almost any cost. By adolescence, he was stealing, joining gangs, homeless, and eating out of garbage bins. He sought revenge for his deprived life through antisocial acts. For most of his analysis, I felt aghast, out of my depth, and unsure how to reach this tortured non-soul. The changes and nuances in the treatment show how Rodney actively distanced himself from his heartache. He tried to speak, act, and be a human being, but he was walking with a ghost who had once been himself, simply trying to get through life. When I was with him, I felt as though I was also on an endless descent into a bottomless abyss.

When I first met Rodney, I stood beside him at the foot of his abyss, peering through the cracks. Chaos descended upon me as he tried to control, dominate, and assert his power over me, splintering and unravelling a sense of stability. His first comment was "You are far too young to be an analyst, and you are an unsuitable choice." Taken aback, I replied, "You have an idea of how an analyst should be, can you tell me more?" He walked out, saying, "I won't be returning to see you," and slamming the door behind him. The following week he phoned me. He was apologetic and requested an analysis. He said, "I have changed my mind. I am happy with your location (i.e. my physical address). You are easy to get to." This statement proved true for other, less conscious, reasons – I was emotionally available, but I was also exploitable, due at the time to my limited experience with someone like him – and he "got to me" frequently, creeping deep inside and stirring up a hornet's nest. My "location" was convenient because, psychically, he believed I would submit to him, and he imagined that by being physically close, he would be able to control me, to keep me within earshot. Already, the duality between reality and his instantaneous perversion of it via *his* alternate reality was emerging. The cracks were widening, and as I peered down the gateway to the abyss, I saw uncharted territory – the dark, seedy world of severe sadomasochism.

I agreed to reduce my fee at his request – a decision I came to regret. Years later, when I increased my fee, he noted in a cold, irritated tone, "You have raised your fee. It is a raise more suited to a more experienced analyst. I don't think your doing this is fair." By implication, he was paying me a reduced fee in order to devalue me. I said to him, "You feel I am exploiting you; you do not think I am of much use or value to you." I was already in the grip of a sadomasochistic transference, although things had not yet escalated, and he was still very controlled. His stone-cold tone and heartless manner stood in contrast to what lay beneath – a tsunami of turmoil crashing inside him. With hindsight, I might have considered his motive for requesting a fee reduction more carefully. I felt pressurized and agreed to take him into my practice without fully realizing he was exploiting me. In fact, I later discovered he could well afford the treatment, and that the actual cost to him was the threat of dependence and of sacrificing his perverse defences in favour of a life-giving alternative. Perhaps he undermined and disabled the frame from the beginning in order to maintain the colossal divide.

Rodney continually disrupted the treatment. He discarded me in the last session of each week, threatening not to return, and reinstated me as his analyst at the start of the next week, so that the weeks felt broken beyond repair. He said on one occasion, "You are the orchestrator, the ruthless head of this scorpion congregation, using me as an instrument to build your territory." I was stunned, still naive to what was unfolding. I replied, "How is it that I am making use of you?" "You seem to think I want to control you," to which he responded, "You think you can collect all your patients to serve you so they never leave you. I am stuck in your venom, so you get richer and richer, and you take everything of mine away." He frightened me, and long after the treatment ended, I recognized that the orchestrator was his powerful and supreme self – an imperious, internalized part-object that reversed his need for dependence in his illusory alternate world.

Rodney triumphed over my perceived vulnerability, regularly noting dismissively, "You are a young and inexperienced therapist" (he was deliberately not referring to me as a psychoanalyst), "just a rookie stumbling in the dark. What gives you the right to think you know something – how foolish, when you know nothing at all." I would try to shield myself and respond as best as I felt I could under threat: "What keeps you here, if you feel you get nothing?" He had the idea that I was easy prey for his predatory self. I wrongly considered him to have feelings and emotions that I later realized he did not have. Throughout the analysis, I felt many of his disowned emotions through my emotional response to him, while he had little feeling – if any at all – indicating the severe split between the part of him that could not know his suffering and the part of him wanting to cross the divide. I felt quite unprepared for a patient whose inner being was as scathing and as vacant as Rodney's. Themes of domination, power, control, and triumph seeped into the room from the start of the analysis. When I registered feeling decimated by him, I pointed out to Rodney that he had devalued me and others, turning me into his garbage. This escalated into raging outbursts, where he became a murderer and I the murdered child. This led to a change in our relationship, as we tumbled into an "inferno", engulfed by paranoia.

Inferno of Attachment

The overwhelming force of Rodney's flammable rage spread rapidly and uncontrollably, like wildfire, as he exploded repeatedly, leaving a trail of disruption. Bullets rained down from all directions. He fought with everyone, causing havoc. While I felt smothered by the fumes of this furnace, searching for a place to hide, Rodney became more violent. Perhaps he did this to destroy and spoil any potential good that might come from analysis. He conned me into thinking that his sharp intellect and dreams were free-associative thought, but in retrospect, I saw this was not the case. His thoughts were fragments, delivered as missiles to others – which was not surprising, given that perversion might be considered a triumphant destruction of the meeting of one mind with another.

As the months and years passed, I tried to keep my mind on an even keel. Residues of intense warfare smouldered. Feelings of dread, disgust, and judgement floated back and forth between us. He would swear at me and say, "You are a bitch, stealing from me with no return, its worthless to me, you are trouble that is all, you give me nothing, you should be a garbage collector." I would reply, "Rodney, if you continue to trample on me like a piece of garbage, I am afraid I will have to stop our sessions." My response was indicative of the need to keep the little boy in the chasm (and myself) alive, while he was annihilating himself through his illusory reality. I never did end the sessions, and he thought my words were weak and empty. I was frequently afraid of him, and this excited him. I could have ended his sessions early when he was abusive, as agreeing to withstand his abuse – without realizing it – was an error. Following his tirades, he would return the next day, apologetic, appearing humble, and saying to me, "I am sorry, you did not deserve that. I thought I wouldn't find you here today," or "I thought I would find the police here today to lock me up." But he was again seducing me to stay – which I did, for many years. I now wonder if Rodney was pretending to be sincere to placate a version of me that, in his mind, would react violently and leave him if we connected authentically with each other. Or, when he apologized to me, was it motivated by an inkling of reality – a momentary realization of what he had done – which then disappeared?

Over time, I was able to talk to Rodney about his criticism of me and others. He was oblivious to what he was doing, and my words had negligible impact. It is exceedingly difficult to know how to talk to a patient who is in a state of oblivion. Eventually, naming the difficulty enabled me to put myself in his position – at a slight distance from my own feelings – imagining what was happening in him. On reflection, I think he was giving me an experience of what he felt like when he was little and with his parents: a state of total disconnection. It was difficult to stay grounded and level-headed while feeling extreme danger. We wrestled for survival for a protracted period. I tried to dodge his bullets; he seemed to duck away from what felt to him like my provocative darts. This warfare excited Rodney, while I continued to feel extreme distress in his presence.

What made this phase of work distinct, in addition to his verbal thrashings and threats, was a severe drought in New South Wales, which resulted in the catastrophic Black Summer bushfires, from which Rodney seemed detached. I, however, was fearful of the spreading fires. The fires inside him, between us, and outside all seemed to combine and challenge my ability to separate his feelings from my own. Analysts are human beings; we move in and out of different states of mind just like our patients, and the environments in which we live affect our ability, sometimes wavering, to keep hold of an inner reflective anchor. I started to notice, towards the end of this phase, that Rodney had very few phantasies and that what he said was invariably concrete and literal. It dawned on me, after months of explosive outbursts, that Rodney had little interest in opening his thoughts up for further exploration. Eventually, I saw that the sessions were about *doing*, not about reflecting on himself or his life. This was an important discovery, as it changed the way I spoke to him. Rodney suddenly began to dream, and although his dreams were at the level of sensation, they were a new mental activity, revealing a change in him.

The Predator's Realm

The explosions died down, and a strange thing happened. The massive projections, communicating his terror, led to a phase of "mutual killing", where it felt as though one of us was going to die. Rodney's paranoia seemed to lessen, although he started to display his cruelty more. I became paranoid at times too. One occasion stands out. Prior to his session, I had a sudden image of Rodney breaking in overnight and my arriving to find him lying in a pool of blood. This was a type of hallucination, which does not surprise me, given the psychic state he was in. I now had murderous feelings towards him too. He shocked me one day when he threatened me: "I will pull out a knife and slit my wrists right here, and then what will you do?" His tone was one of savage delight, although he was, of course, not aware of my fantasy prior to his entry. He was playing with me. I felt frightened, like a deer in headlights, confused by the thrill on his face. When he noticed my fear as he gauged my reaction to him, he smiled ruthlessly. I had no idea what to say to him and simply replied, "You seem to think it will kill me more than hurt you, which makes you very happy." (I was at a complete loss as to how to respond to my own fear.) He then added with glee, "I often think about an axe murderer, and I like the thought of blood splattered on walls." I replied, "killing, murdering is thrilling, and it feels like revenge for the hurt hidden behind." He appeared smug and remained untouched by my comment, laughing at me. He was likely projecting his disowned suffering onto me, severing any feeling reminiscent of the agony and torture he had felt as a child. Today, I might have said to him that if he slit his wrists now, I would call an ambulance – to bring external reality to bear in contrast to his perverse reality, into which he constantly fled.

On a different occasion, Rodney had a dream that he was in the Everglades. He said, "Crocodiles surrounded me. As I moved towards the swamp, I saw two huge green eyes. They dragged me underwater as the crocodiles sank their teeth into me." I said, "Do you have any thoughts? That sounds brutal and savage – a vicious attack." He had none and replied, "My father was very cruel and abusive toward me; he was a beast, a real bastard. I think the dream is about my father." Although I tried to prompt him to pay attention to his dream and see where it took him, he could not elaborate or personalize the dream. I was unsure who the crocodile was or how to keep abreast of these intense, savage feelings of aggression, as well as my own feeling of drowning. I suspect that the crocodile was the part of himself that devoured and decimated him, trying to slice me up. On the other hand, he may also have been trying to show me, through his dreams, that he had nowhere safe to go – that he needed to move further away, as he felt he would be devoured. I felt like an antelope at a swamp, nervously looking around, in fear of imminent attack or death, alert for a predator to pounce. I imagined that if I ended Rodney's treatment, he would threaten to kill me and attempt to destroy my career. Fleetingly, it felt as though he were a giant monster. Again, the duality of the colossal divide – evident as he tried to kill himself, end the analysis, and by extension murder me – threatened to overwhelm us. I would speak, and he would swat me away like a nuisance fly, easily forgotten. He did this by outsmarting me, showing that he had a better idea of why things had happened as they did, and when I made comments, he would say, "I have thought of that already, it's old news." Although he was a hollow man, he could reveal his inner world through dreams, indicating some life inside him, but the way he showed me this life was by inducing me to "live out" his problem. It was exceedingly difficult to feel empathy for someone so very cruel, and this was a paradox. However, the fact that I remained committed to his treatment was revealing of his fleeting determination to become alive.

In a different dream, he said, "I was dismembering people; I hid their body parts so I could conceal the murder." I responded by enquiring, "You felt worried about what you had done in the dream?" He replied, "No. I was worried that I would be caught by the police for committing the crime." He felt no guilt about what he had done to himself, to me, or to others, as he had little conscience and limited, if any, capacity for concern. However, I carried his concern, reflecting the duality of his split mind. He was possibly dreaming about the murder he was committing over and over again in the room – which I did not realize at the time – and about his own murder.

Rodney had lacerations on his legs, which he would proudly show off. He thought he could rouse my excitement by showing me how he abused himself. Destructiveness protected him from his experience of suffering. By contrast, I suffered deeply, as I embodied his tortured self. I did not always realize that I was afraid of him, because I was under his full control. On one occasion, he reported that he had watched *National Geographic*: "a leopard tore at the anus of a vulnerable fox. The fox was left with a gaping hole. The ranger, a strange man, callously said, 'It's life, animal's instinct, born to kill.'" His tone was cool, calm, and collected. I replied, "The fox was left to suffer, and the ranger had a duty of care but was cruel and callous. You know about being left to suffer alone, with someone who despises you." With the benefit of hindsight, I realized that Rodney had turned me into a killer. Although he was the real murderer, he could not bear these feelings. A ruthless aspect always tore into others – into me. Interestingly, throughout the treatment, we both lost sight of the vital moments when he was alive, largely because he quickly perverted them.

He felt his mother had owned and destroyed him, that he was an object for her use, so he murdered himself and tried to murder me. With hindsight, I might have interpreted to him what he was doing to me while he was in identification with his mother. However, if I had interpreted it, it might simply have been a conceptual idea for him. I did not understand then that killing might have been his way of taking possession of me (or others). Rodney was showing me what it was like to be crushed, bruised, and terrorized – not to have a real relationship, to live in emptiness, nowhere, to not be anything or anyone, with nothing to build upon, no trustworthy good experiences. He was showing me that people were dangerous and had to be avoided.

Rodney was taking revenge on me, yet underneath he was really an empty, hopeless person – charred remains smouldering.

It took nine years of analysis for me to understand that Rodney was an expert at using words, but not for communication purposes. Words were weapons – missiles launched to ensnare, control, or seduce me – rather than to convey conceptual or emotional meaning. His words were very impersonal. He used them to convey pain that he could not feel. For the first time during the treatment, I felt some empathy for him, which was a meaningful change. This realization was important as up until this point, I had listened closely to the content of what he said to me, not to how and why he was destroying meaning. In the next phase, I was able to tune into other senses, rather than only listening to his words, and this led to a development in the analysis.

Chains of Desire

After the prolonged warfare, the heat died down and his paranoia subsided. The atmosphere changed as Rodney sexualized his hostility. When entering the consulting room, he smiled seductively. I felt uncomfortable – chained. When he lay down on my couch, I felt relief at no longer feeling his eyes on me. He also became very sexually charged outside the room. If violence was inflicted upon him, he was sexually excited or detached – but never anxious – and it reminded him of times in the past when he had used the sexualization of his body to feel something, anything at all. He relayed stories of lust and passion, imbued with punishment.

Rodney's phantasies centred around the pleasure he derived from tying women up and engaging in perverse sexual acts. His invention was that I, in addition to primarily being a murderer, was also a seducer, and since his thoughts were facts, they felt very real to him. I thought he was self-medicating to numb the shattering experiences of neglect.

His seductive manner resulted in my feeling repulsed by him, this being suggestive of his repulsion towards his mother when he was a small child. I spoke to Rodney about his confusion. Erotic feelings, love, and sex were all mixed up for him. I saw the little boy in the room craving my attention, not the grown man. He was not able to discriminate between having a sexual life with a partner and an intimate emotional experience with his analyst. When he was in a highly eroticized state, and my feeling of isolation from him was increasing, I said, "You are starving and are looking for a mother with whom to feel safe." On many occasions, I needed to show him that I was not who he imagined me to be in his fantasy (i.e. not his father, nor his lover, nor his mother, nor his slave). As I pointed out his tremendous confusion, a meaningful change occurred. For the first time, I spoke directly to him about his delusional reality, while simultaneously feeling a stronger understanding of how isolated he had been as a child, needing to invent this alternate reality. Banished feelings returned, from an unlived life in a frozen wasteland, where warmth was as distant as a sun that never rises.

Until that moment, I had thought that Rodney had two vital imperatives: not to depend on an object, and not to feel inferior to his object. I now understood that it was not a case of his feeling inferior to his objects; it was a much more severe problem – there were no objects to depend on. If he were to enter the human world, it would have the effect of bringing him face to face with the fact that he had never entered it before, and that he felt he did not deserve to be part of it. He lacked an entitlement to live; he had no home inside himself and no roots. Not existing at all protected him from this catastrophic potential discovery, reinforcing his need for a delusional reality. To counter this objectless world, he would make things happen to prop up his alternate reality. In the first phase, he did this by usurping power; in the second, by inflicting cruelty; thereafter, by killing; and in this phase of treatment, it was by hostile seduction. Navigating the "catastrophic chasms" was a turning point, as when we spoke to each other, the quality, tone, and atmosphere of the dialogue had changed. It no longer felt like we were at the front lines of a war or in the eye of the storm. In place of this, I was now able to notice the side of the colossal divide that had been permanently sealed off – namely, his vulnerability.

Navigating Catastrophic Chasms

Rodney began to soften once I was in the abyss with the intention of exploring it. This was a tremendous change, as it felt like we had plummeted into a chasm of total darkness. I felt interested (which was also a new feeling) in the fact that Rodney was now starting to have dreams revealing a sense of isolation. He still created distance between us, which is why I continued to feel lonely. Unexpectedly, Rodney reported watching a television series called *Alone* – a documentary about how individuals survive in the wilderness using minimal survival equipment. Participants are sent into the Arctic to build shelter and find food amidst wolves and foxes. In a monotone, Rodney described how they survived the loneliness: "They chopped down trees to make chairs, ate plants, tracked animals, made earrings out of stone – all alone, happy to be far away from civilization." I said, "You are describing yourself reassuring a younger you that it's safest to never need me or others." He replied, "In the 'wilderness' those foxes and wolves are devils – you must watch out all the time, they creep up on you, and without a machete, you are a goner. I would shoot them at once if I saw them – make them die a slow death." He continued, "One day I will be able to tolerate isolation. It has not driven me mad, and it will never. It's my friend," he chuckled dismissively. This statement was Rodney's altered reality – in actual reality, humans are social beings, and if somebody is put in isolation without human contact, they will not survive. However, in his perverse reality, he could invent anything he liked, and he believed that he had the power to tolerate exile.

As we delved further into the abyss, Rodney had another dream. He said, "I was living in an igloo in the Arctic. A wild storm arrived, there was no visibility, and it was icy cold. The igloo gave way and collapsed. It was no longer a safe shelter. African thugs came to eat me, tearing at my flesh." I think he worried that I (his South African analyst) could not tolerate the pressure of his alternate reality. The dream seemed to describe his paranoia. The domination of the cold, the ice, and the life-threatening conditions in this Arctic wilderness showed how dangerous any attempt to exist was, and how unsafe he still felt. When I tried to reach his heart and soul and said, "What a terrifying place to live, and then to die so cruelly," he spat at me like a venomous snake and said, "Shut the f – up, I don't want to know." I was a figure who he loathed largely because he viewed me as a person with emotional resources – someone who could abandon or replace him. Again, as he felt my presence, he simultaneously pushed me away.

Rodney was perplexed, and for the first time since we met, I was a mystery to him, in that I no longer fit into his predefined mould of who he thought I should be – namely, someone who would kill him. As a result, he did not know what to do with me. Although he still warded off any feelings of pain, I thought he had begun to trust me, spending a little more time than before in actual reality (in the abyss). This change was the catalyst for the transition into the final phase of treatment: the "illusion of Born Free".

Taking Bait: The Illusion of "Born Free"

I felt closer to Rodney, which seemed to make way for a freedom I had not experienced with him before. Until this point, he needed to obliterate anything or anyone who roused feelings of dependency, and I thought this was beginning to change. I felt I could begin to find my own thoughts, separate from his. Rodney introduced a dream, "Born Free", about a documentary set in Africa, my country of birth. His dreams were still factual and emotionless. He said, in an animated manner I had not seen before,

> An orphaned lion cub, Elsa, was adopted by a couple who took responsibility for dealing with dangerous animals that chronically threatened humans. This couple killed Elsa's parents – man-eating lions – leaving three orphaned cubs. One was Elsa who the couple adopted. The other two cubs were sent to the zoo. But Elsa became an adult, and they released her into the wild.

He continued to give other details. After an extended period, I said, "A small, orphaned child taken in and cared for . . . remind you of anyone?" I had thought this dream illustrated that Rodney could depend on me and separate from me. Perhaps there were moments of this; however, he mostly still starved himself and had no idea how to connect. On reflection, perhaps I was lured into a trap. The "Born Free" dream might have simply been a respite from his alternate reality, in which I was a predator or killer.

As I listened to the story about Elsa, I responded concretely, seduced by Rodney into thinking that the cub image expressed infantile aspects of his dependent self, which I thought offered hope that Rodney could free himself from the primal threat. I thought the "cub" – that is, baby

Rodney – felt cared for by me. He said, "I feel you have taken me in. I am desolate, I have nothing, and I expected you to leave me – who wants a dying dog?" I underlined his words and asked curiously, "A dying dog?" He stated, "I feel you have helped me realize how crazy mad I feel and how often." This statement felt like an expression of a truth. In hindsight, I had many thoughts about the exchange. Perhaps it was seductive, in that he had made me into someone benign – a good and loving parent who had rescued him. It might also have been an escape into thinking he was "free", rebirthed by the analysis, and no longer captive to his delusions.

Alternatively, it might not have been an escape, but rather a comment expressing that he felt my interest in him. If it was the latter, the perversion prevented him from staying with this awareness for more than a moment, as he would inevitably revert to his paranoid world. Since the man-eating lions were banished temporarily from the analysis, it is possible that he got rid of his aggression and hatred. Rodney had been killed in infancy and had become a "killer" as a way of concealing self-hatred. A further possibility is that his dream might have been a manic defence, an illusion, and a response to his fear that I had seen who he really was. In mania, the person suddenly becomes convinced that he has the power to magically make things better. He had in fact felt more contained after he dreamed about the *Alone* TV series.

The lion cub (Rodney) had to make a transition from human captivity (his inner sadist) and return to the fearsome wild (freedom from the sadist). The unbridgeable great divide between these opposites – captivity and freedom – left a gaping hole, a split between his captive self and the freedom he might one day finally create for himself. I felt I was in an "analytic captivity" for many years. One of us always felt like a caged, subhuman, or non-human being. Rodney, as captor, held his grip on reality.

Rodney ended treatment suddenly at the point when he was no longer able to control or seduce me into going along with his alternate reality. He realized that I had seen through the falsehood that he had found "freedom", and my observation was terrifying for him. I had wondered if he had begun to feel a bit closer to me and that the psychosis in him might have convinced him that he was cured. He certainly was not free, but the major turning point in this phase may have been his wish to believe that he was free – this being the altered reality. I have also wondered whether he possibly tried to prove he was living in reality, not in a delusional world, to protect me from the horror of who he was and to conceal the alternate world he lived in, making me believe he was free. "Born Free" is a world Rodney had never existed in.

It was with surprise that I had the thought that he may have unconsciously and psychotically manufactured an image of progress, change, and development. He also created his two worlds in the analysis – an analysis that was progressing (in his alternate reality), alongside a dead, toxic infant within him (actual reality) who battled to cope with the analysis.

His psychosis was powerful and demolished his true experience. It made him believe that by telling me about his sense of aloneness and his experience of the wilderness, and by confiding in me more, that I would despise him, kick him out, and repeat the catastrophe of his infancy. Briefly during the analysis, he could experience himself and me as human. However, he always returned to the wild in which he needed to be the king – a supreme being, the ever-powerful one – to manage his ongoing, unfelt terrors. These speculations were all unknowns, based on what we had lived through together.

The fact that I could speculate at this point, unlike the suffocating chaos of the furnace in the "predator's realm", demonstrates that Rodney did make a significant impression on me, and perhaps it is indicative that his contact with me might have made a lasting impact on him too.

Despite all his efforts to fool me, to do what he had always done, by the end of the analysis he had only partially succeeded. The analysis had opened a chasm in which he felt there was nothing – that he was nothing, and that his analysis was nothing – though I do not think this was the reality, as I feel it gave him a great deal in terms of an experience of a non-perverse world. Thus, the period before he ended the analysis was a very scary time for him. The analysis offered him a clear picture of what he was up against (in himself), what he was hoping for, and how impossible it felt to him. I did not get a chance to work on the "nothingness" that had always been with him in his life.

In our last session, he said to me, "Your heart was in the right place," and I suspect it was perhaps the very thing that kept him returning to see me, again reflecting the duality of his inner predicament. This comment of his has lingered, perhaps because he at least had a sense of someone with a heart. He watched me very closely, and I think by the end of the analysis he knew I was not his father and that I was not going to kill him. Before we met, his world and the people in it felt heartless to him. I do not feel this comment was seductive. He saw, however, that he was unable to be a person, that he did not have a heart, or that his heart was battered, and so he could not go into the nothingness.

Further Reflections on the Transference–Countertransference Crossfire

When I first became aware of Rodney's wish to control and have power over me, as he launched into the "inferno of attachment", I noticed that if he created space between us, he did so by increasing his distance. If I was in a relaxed state, which I seldom was, then he sensed my efforts to consult my analytic mind. Perhaps he felt that my internal dialogue was a way of excluding him and threatening his existence. I had aligned myself with the part of him that was alive, but this part was under constant surveillance and constantly annihilated. As the "predator's realm" unfolded, I noted intense and unrelenting distress and fear while in his presence, but it was difficult to step aside from my emotions and view the experiences from a distance. Thinking back, it might have been helpful to get Rodney to do more of the work. I might have enquired more, helping him expand his comments even slightly. Perhaps it might have created some space for me to think, and for him to feel things. I cannot know. Interestingly, when I allowed him to control me, I became an object devoid of humanity. On occasions when I did point out my understanding of his subjective experience with my objective view of the situation, he screamed at me saying, "You are an idiot, shut up and listen to me." Pointing out my understanding provoked either violence, withdrawal, or both. On many occasions I was obedient, he ordered me around and I complied like the terrorized small child he once was, and whom he despised. I regret this, but on many occasions, I also did stand up to him.

His outbursts were vacuous, and it felt as if I were drifting aimlessly like floating debris left after a storm surge. I blamed myself for his fury and his accusations about my "incompetence". This self-blame was irrational, as it was the result of Rodney's masochistic projection. He said, "If my father heard a pin drop, after demanding I be silent for being disobedient, he would f – me up and attack me like a pack of wolves." His statement was reminiscent of the nightmares he had reported during the analysis. Perhaps he felt he was receiving a whipping from a tyrant father who he felt was me. He tried to erase me, and when I was not "obedient", he felt I was erasing him. In hindsight, aligning myself with Rodney's non-psychotic part, while drawing his attention to his psychotic part, might have been more of an opening to the colossal divide within.

A distinct change in the "chains of desire" was my increased feeling of isolation and the fall into the chasm. Finally, I felt I was standing close to the little boy and his unfelt pain. He employed sadomasochistic shields, but he still had to carry a dead carcass around to get through life. He was internally a hollow echo, and I could bear witness to it now. While navigating "catastrophic chasms", the solitude of an infinite sinkhole appeared, where Rodney felt he was nowhere. Faced with a tumultuous, turbulent, deafening, chaotic echo of his childhood trauma, movement and change were occurring. He could not reflect upon his experiences, neither could he entertain another point of view. I too often battled to name my own thoughts and feelings, while simultaneously communicating to Rodney my understanding of his point of view. When I tried to do this, an internal avalanche crashed down on me, fracturing and crushing the source of my thoughts. His paranoia was infectious, and it often (though not always) destroyed the potential for a sustainable lifeline between him and me, although this changed in "the illusion of Born Free". In spite of my increasing understanding of his condition, towards the end of this phase I had irrational and likely delusional thoughts that, if I ended his treatment by ending my career as a psychoanalyst, I would no longer be what I felt myself to be – namely, enslaved by him (this was an aspect of the sadomasochism). Rodney's irrational fantasy was a powerful projection of suicidal anxiety and despair lodged in me. It was as though the two of us were to be dragged into a comatose non-existence together. The greatest threat was to come alive and for him and me to relate as two separate individuals.

When Rodney ended the analysis, I was relieved to get rid of the agony of insufferable tyranny and to regain my mind. His decision was a response to my arriving cumulatively at knowledge about him that nobody had ever recognized before. He no longer desired to prey on me. This corresponded with my firming up and not allowing myself to be drawn into his well-established sadomasochistic battlefield. Once I firmed up, he was unsure of what to make of me and he felt profoundly anxious. I was in fact closer to his isolation than I had been previously, which I believe contributed to his sudden departure from the analysis. Although I tried to talk with him about how and why he had arrived at this decision, it did not affect him. Despite the years together, he was still unable to form a human relationship.

By the end, Rodney's sadomasochistic and perverse defences were proving ineffective. He no longer felt as excited as he had before. When he fought and disconnected from me, it did not arouse him as it used to. Sadomasochists always have the object within reach and at arm's-length. When I ceased to engage in the arm's length battle, Rodney started to feel desolate and bleak – feelings he had not felt while engaged in the masochistic fighting. The gap between the two realities – between the sadomasochistic gameplay and the desolation – was enormous. The threat of mutual eradication, however, still felt safer than the threat of human contact, which was disarming. The threat of what lay underneath his need to eradicate himself and me – namely, "nothingness" – was so unbearable that it caused him to quit the analysis. Rodney used massive expulsive projective identification. This overwhelming, massive projection into me had everything to do with the horror and destruction of his existence. For a whole year after the termination of his analysis, I felt relief at not seeing him. I did not have a way of protecting myself from Rodney's horror. If it happened again, I like to think I would catch it sooner, protect myself earlier, and keep myself from being so vulnerable to it.

Conclusion

The many years of Rodney's analysis were like a long trek. I felt I had to find a way to navigate to the bitter end of his emptiness and hopelessness, finally arriving at the realization that he believed nothing was alive within him and that he felt he was worth nothing. The irony was that the very act of doing this did create some life – not historical life, past life, or restoration, as all of that was destroyed in infancy. This "trek" was something new, as I was willing, despite my own limitations, to try to reach him, to witness what had happened to his soul, and I was willing to go the many miles with him to the arid, scorched place where he began to discover there was nothing. What were we to do then? I learned many years later that we must accept there is nothing – if that is the truth – but we have each other, and that is the new experience. Rodney ended the analysis prematurely, as his isolation was too unbearable. I was not able to show him that I could go to the deepest part of the chasm with him, to the heart of his agony, and he was not able to trust me enough to do so either.

Notes

1 See L. Shengold, "Trauma, Soul Murder, and Change", *Psychoanalytic Quarterly*, 80.1 (2011), pp. 121–38; P. Williams, "Isolation", *Psychoanalytic Dialogues*, 29.1 (2019), pp. 1–12.
   Shengold describes soul murder as the apparently wilful abuse and neglect of children by adults that is of sufficient intensity and frequency to profoundly inhibit the children's subsequent emotional development. Their motivating unconscious fantasy life is dominated by what has been done to them in the past. As a result, they feel compelled to repeat the same patterns of cruelty, violence, neglect, and sexual overstimulation.
   For Williams (2019) if the core self is forced to give itself up entirely to an invasive object, the ego collapses and the self dies, leaving a barely functioning body. When the will to live is so compromised, what survives lies at constant risk of self-harm. In practices of torture, the core self is exposed and mutilated before being destroyed. Psychic death occurs when the subject merges with the object, "becoming" the torturer or murderer. The phenomenon of the concentration camp "Muselmann" is a graphic example of destruction of the core self.
2 "He" or "him" refers also to "she" and "her", "them" and "they".

4 The Psychoanalyst's Awareness of Climate Trauma in the Clinical Situation[1]

Sally Weintrobe

Monica

A colleague asked if I could see a woman for a consultation with a view to possible analysis. Monica, a woman in her late twenties, was seeking psychoanalytic help with a problem. She wanted a baby, but she felt that having one would be irresponsible given the climate emergency. She stipulated that she wanted an analyst who is "climate aware". I was not able to see Monica at the time, and this is all I know about her request for help.

Monica's subjectivity is uniquely her own, but her dilemma is now a common one. In September 2021, *The Lancet* published the results of a global survey involving 10,000 young people aged between 16 and 25 from countries across the world, including those in the global South.[2] They were asked what they thought and felt about climate change. Seventy-seven percent said they found the future frightening, 40% expressed serious doubts about whether to have children,[3] 60% said they no longer trusted their governments to tell them the truth about the climate crisis, and 50% said they could not talk to their parents about climate change, as their parents did not want to know. Women's decision not to have children was raised on BBC Radio 4 during the COP26 negotiations in Glasgow in November 2021. One woman said that, given the climate crisis, she could no longer imagine offering a baby or children an adequate sense of home, safety, future prospects, and ordinary happiness. She said she was distressed both by how she would cope with being a mother and by the knowledge that the children would suffer. Meanwhile, world leaders at COP26 failed to offer even non-binding pledges to keep global warming below 1.5 degrees Celsius.[4]

Monica sought analytic help in 2019, the year Extinction Rebellion was declared, and there was a 20-fold increase in the use of the term "climate crisis".[5] The word "crisis" has multiple meanings. In relation to climate, it seems to be mostly applied as a diagnostic term, in the way it is used in medicine: the critical point in a serious disease process that determines whether the outcome will be recovery or fatality. A crisis is a tipping point reached. By 2019, the terms "climate anxiety" and "climate trauma" were

DOI: 10.4324/9781003634959-6

also in common usage. Climate anxiety is used to refer to a wide range of different feelings and reactions that accompany seeing climate reality clearly, without disavowal. These feelings include dread, rage, grief, shame, guilt, despair and, at times, a sense of overwhelm. Experiencing these sorts of feelings is widely considered to be evidence of a mentally healthy reaction to seeing the extent of the damage to the planetary systems that life depends on.[6] Climate trauma has elements of both post- and pre-traumatic stress disorder. In both cases, the world no longer feels safe. Another form of climate trauma identified is moral injury – the awareness that one is both victim and perpetrator in a way of living that violates one's sense of what is right and decent.[7] With moral injury, the world no longer feels moral, and it has been found to occur in cultures that are casual about killing.[8] In my work on climate, I have situated the experience of such feelings as contributing to emerging from what I have described as the "climate bubble", an organized, collective psychic retreat from reality,[9] promoted by a culture of uncare that actively invites our collusion.[10]

"Climate awareness" is a term that has arisen to describe people struggling to emerge from the climate bubble and expose themselves to often unbearable feelings – unbearable because of inaction in addressing climate change and the resulting accelerating damage.[11] The damage is also to the mind's capacity to contain knowledge about this level of harm. Climate awareness is not solely intellectual awareness that there is a climate emergency. It refers to a state of being open to suffering the feelings that accompany gaining climate knowledge. Returning to Monica, her presenting problem involves a mother, a baby, a climate emergency, and someone who is – as well as someone who is not – potentially climate aware. Considering her problem dynamically, through a psychoanalytic lens, might the climate emergency concern the physical climate, Monica's human climate, or perhaps both? Might the unaware object be Monica's potential analyst, Monica herself, her archaic objects, culture at large, or our politicians? Any clarification of questions like these could only hopefully emerge through therapeutic work over time. We would expect to find elements of external and internal reality interwoven in Monica's material, particularly if the underlying terrain is trauma. We know from trauma research that external trauma likely reignites personal trauma, and personal trauma likely colours how we experience traumata with external causes.

Attempting to untangle such interwoven strands requires a disciplined approach towards clinical material – one that includes tolerating uncertainty or not knowing, and feeling lost at times. It involves triangulating: that constant refocusing of the psychic lens, shifting from being in the therapeutic couple to looking on at it. This triangulation helps us better understand collusions and enactments. For example, too readily agreeing with Monica when she says that the climate emergency is planetary may be colluding with her in avoiding potentially more difficult personal strands, as things that are "true enough" are often used for projective identification.

Alternatively, we may treat the anxiety she expresses about the state of the planet as a mere metaphor, when it is actually the point of urgency in the session. Monica may go along with us for the sake of a quieter life (for instance, not standing up to her analyst when she actually disagrees with her), or because she feels disempowered and not entitled, or for other reasons.

This kind of collusion has been recognized and described in accounts of analyses with white analysts and black patients, where race just never comes up, or it is swiftly redefined as representing something else. Or with female patients and male analysts who do not understand the pain of gender discrimination. I believe the essential dynamic is no different in the case of patients voicing anxiety about climate and the state of the environment. In the 1989 film *Sex, Lies, and Videotape*, a patient tells her therapist she is concerned about all the rubbish. Mounds and mounds of rubbish are piling up. Where does it all go? The therapist says that what she is really talking about is the state of her marriage. She may well have been, but to me, the therapist in the film does not empathically engage with her attempt to understand the world as she sees it.

How each person in the analytic couple sees and represents our world today is bound to affect how the analysis will unfold. In *Psychological Roots of the Climate Crisis*, I contrast, very broadly, two different representations of the world, as Planet La La and Planet Earth. Both are psychic positions. To live on Planet La La is to live inside the climate bubble – an organized collective psychic retreat from climate reality. To live on Planet Earth is to struggle to face climate reality. Most of us oscillate between these two psychic positions. I certainly do. Reality is a contested idea, and different disciplines – philosophy and the natural sciences – approach it somewhat differently. Freud essentially saw reality as that which, by being there, stands in the way of, impedes, or disrupts wishful phantasy.[12] Freud's definition serves us well when thinking about climate reality.

We have a climate crisis because people in power have insisted that wishful phantasy triumph over reality. We saw this insistence during the COP26 deliberations. The climate scientist Kevin Anderson put it succinctly, saying that political leaders were caught up in a "shiny green glow" at Glasgow.[13] Part of this shiny green glow, I believe, is a sense of triumph at getting away with finding omnipotent ways to avoid limits and loss – for example, redefining real zero as net zero, or agreeing to phase out subsidies not for fossil fuels in general, but only for "inefficient" fossil fuels. In these ways, the perverse wriggle room to evade reality is maintained. Anderson noted that in the room where climate scientists were discussing the truly grim current state of the cryosphere, with Arctic melt and methane release, it was a case of scientists talking to themselves – as usual – as they watched planetary tipping points being reached. Young people gathered in their thousands outside the official venue at COP26. They know the issue is a matter of life or death. I was moved to hear Elizabeth Wathuti, a young climate activist from Kenya, chosen as one of a small number of youths to address world leaders in Glasgow, begin by saying, "I've thought long about what words might move you. My truth will only land if you have the grace to listen."

The Setting

My argument is that how the analyst sees – and feels about – the world today will significantly impact the setting she is able to provide. The analytic clinical setting is designed to illuminate transference, countertransference, and resistance phenomena. Put another way, it provides conditions that can help the patients separate their phantasies of objects from their view of those objects as real (having substance in external reality).[14] For example, the patient suddenly sees that they have an idealized view of the analyst, and that this is separate from their view of the analyst as real. Bringing the setting into the picture, Strachey (1969) believes such moments of insight have the potential to shift the way the world is seen, referring to interpretations that facilitate such moments as "mutative interpretations".[15]

Returning to Monica, let us suppose that analysis could help her recognize what belonged to her personally. That way, she would be better able to disentangle her personal angst from her climate angst. She may, for example, come to understand through analytic work that she can experience her analyst – her mother, herself – as narcissistically entitled and cruelly indifferent to the needs of a baby. She will, hopefully, then be better able to tease this out from a perhaps broadly similar view she also holds of her government – namely, that her government is inflated with Exceptionalism and cruelly indifferent to the fact that its policies, and the ideology driving them, are inexorably leading to the climate catastrophe already directly affecting her generation. Being able to make that kind of distinction would hopefully enable Monica to make better decisions about her life.

Strachey's argument takes into account the necessity of taking *both* phantasy *and* external reality into account. After all, moments of insight are those in which both can be seen and compared side by side. Here, I turn to a tendency that would seem to have crept into analytic thinking and framing. This is the position that says, "Leave politics, culture, and ecology outside the door. They are not part of the psychoanalyst's domain." This position fails to take into account the extent to which thinking – and indeed the quality of thinking – is itself influenced by political and cultural framing. It is by now well established that climate denial is largely culturally framed and maintained.[16] Indeed, the idea that I, uniquely, as an individual, can entirely avoid being influenced by the current dominant and pervasive culture is implausible – particularly as this culture promotes perverse arguments, hubris, and the notion of being an exception.[17] Nevertheless, it can be difficult and humbling to recognize the extent to which one has been influenced, and it is painful to realize one has colluded with such perverse arguments.[18] I believe that to deny that culture, politics, and ecology fall within the psychoanalyst's domain is to position the analytic dyad as existing in the equivalent of a hermetically sealed petri dish. It creates a blind spot that refuses to see parts of reality as being represented in the mind. The danger is that the analytic setting itself may become a form of psychic retreat from reality.

After all, we apprehend external reality through representing it psychically (both consciously and unconsciously). The analytic setting exists in the mind of the analyst – there is nothing external when it comes to the setting. Creating a negative hallucination that *apparently* ablates parts of external reality is achieved through disavowal, which can block communication with the patient, who may be left thinking, "What planet are you on? Planet La La?" We need to be alert to the dangers of presenting a cloth ear and a blind eye to our patients' anxieties, which may involve a maelstrom of external as well as internal factors. Being alert may require us to re-emphasize to ourselves the importance of external reality in all its manifestations, while also not losing sight of the power of wishful phantasy to shape how internal reality is experienced.

It is also important to pay attention to the nature of the initial and evolving contract established between patient and analyst. Some contracts themselves would seem to promote the establishment of a collusive pact between patient and analyst to – largely unconsciously – agree to avoid certain problematic areas such as social class, race, gender, and environmental concerns. We might speculate that Monica, for instance, is saying that, as part of her initial contract, she does not want to have to contend with active climate avoidance; *that* will not help her distinguish phantasy from fact regarding a very real problem she is facing.

The issue of the contract is complex – too complex to explore here – and there are usually mismatched expectations at the outset of any analysis, the patient perhaps seeking an idealized "as if" solution to their psychic pain, while the analyst must relate to the patient where they are, hoping that greater realism will prevail in time. However, the reverse can also apply: the patient may initially encounter an analyst caught up in climate denial or in the denial of other largely institutionalized structural social issues.

Isolationism

Climate denial takes a myriad of forms. Isolationism is just one. The thinking is like this: yes, granted, the climate crisis is real and very serious, but more pressing crises are here right now – the war in Ukraine, the economy, racism, gender inequality, and the crisis of democracy.

Climate must wait. With isolationism, all these crises are thought of as separate from each other; moreover, addressing climate is always placed at the end of the "to-do" list. However, addressing climate change must be the top priority. We have, at best, less than 10 years to drastically reduce carbon emissions, or life on earth will be severely compromised. Isolationism dulls the will to act and prevents us from understanding that the current emerging crises are intertwined and stem from an uncaring, purely extractive and exploitative mindset taking charge. Indeed, awareness that politics, culture, and ecology are deeply connected is part of being "climate aware". Being climate aware involves undoing the mental fracturing of isolationism and recognizing that the problem of climate is, at its root and in essence, a psychological one. It involves a mindset that attacks care and care's restraining hand. Politics in this sense is not party political, although it may appear so at the surface level. I believe isolationism has – unwittingly – played a part in maintaining the position that politics, culture, and ecology can be left outside the consulting room. However, there is another potential barrier for the psychoanalyst seeking to broaden her conceptual field.

Method

This barrier is methodological. One potential problem for psychoanalysts whose disciplined approach to psychoanalytic data has been honed through working with individual patients on the couch (I am one of these) is the anxiety that if the wider social context is now to be included as part of our setting, we may lose our disciplined approach and stray into "wild analysis". While this anxiety can be well founded, I believe it is based on a misplaced preconception: that it is legitimate to simply apply findings from the study of the individual to the study of the social (or from the social to the individual) in a wholesale, "blanket" way. It seems to me that what *is* legitimate is to take it seriously when we observe the same – not identical, but broadly and recognizably similar – patterns of defences occurring at these different levels. They can and do recur at different levels, like a fractal pattern repeating at different scales. Examples are the way in which disavowal of climate reality is found at the level of the individual psyche, social groups, and the political process, such as at COP26. The misapplication – the error – is to conclude that the underlying reasons for the defensive organization will be the same at these different levels. Also, crucially, trying to understand a pattern of denial at, say, the social level will require bringing in other disciplines. Psychoanalytic understanding applied at the social level has to address a different data set, and to do that, it must find its place alongside other disciplines.

It is insufficient to simply exclude politics, culture, and ecology from the setting as a quasi-"solution" to new and necessary theoretical work and methodological discomfort. Application requires conceptual work, which can lead us to broaden our understanding of the setting in productive ways. We have seen this with COVID-19. A point made in a paper by Caroline Polmear was that a changed conception of the setting, brought about by Zoom analysis, could shine a light on corners where patient and analyst had colluded to keep out various painful areas. I believe climate has been one of these avoided, painful areas for many analysts. For the analyst to take the climate crisis seriously – understanding that, because action was not taken earlier, it is now literally a matter of life and death – can expose her to heartbreak, and working through that can at times feel unbearable. Irma Brenman Pick reflected in a personal communication (2022) that the depressive position is not adequate to describe working through the current unfolding realities, and that we need a tragic position. She said, "The tragedy is that the defences used, like denial to avoid the pain of knowledge, in fact increase and intensify the very pain that is being avoided". This could not be more apposite for our times. Inside the climate bubble, people have minimized awareness of the damage and suffering they contribute to causing, with the result that both have accumulated. By the time the climate bubble bursts, some of the damage is irreparable, overwhelming to grasp, or both. The tragic position deals with the pain of trying to bear the unbearable. The unbearable includes the whole range of our complex feelings – indeed, anguish – about what is irreparable. We are confronted with intergenerational conversations that, if honestly faced, would expose that the young are still being awarded no entitlement to a viable and trustworthy world. And, of course, climate reality will feel different to young people and to children, who are more directly affected.

Fragility in the Face of the Unbearable

As analysts, we are familiar with the unavoidable anxieties that stem from our human fragility. However, in today's modern world, the climate crisis is driving a mental health crisis, as we struggle – often unsuccessfully – to contain new levels of existential threat. This threat is often posed as one to physical survival alone, leaving out the profound distress being caused to people – including children – who are suffering a breakdown in trust in those meant to take care of them. We know, as psychoanalysts, that in order to love, we need to be loved. Part of what feels unbearable today is seeing so many young people realizing that their governments' plan is to sacrifice them by not acting on climate change. How do we expect children and young people to be able to weather this kind of emotional storm?

Returning to Monica, let us suppose that she may have emerged – or begun to emerge – from the climate bubble. In that position, she faces – as we all do – the series of shocks that accompany emergence from a collective psychic retreat from reality. The shock of this emergence will likely ripple out in waves. Finding herself on Planet Earth, she will likely be in a state of considerable anguish, contemplating her present and future choices – one of which is whether to try for a baby, a life-changing event that will evoke her wish to provide a containing physical and psychic environment in which her baby can grow and thrive. It is well known that mothers require support to manage this immense task. Monica will need a stable climate to raise her baby. And here is the tragedy. Because of industry-sown denialism and our widespread collusion, the physical climate is no longer temperate but is instead weirding, increasingly unheimlich, and rapidly becoming dangerous to life itself. So too is the emotional climate, as these two climates are interdependent. This is the world that Monica faces as she thinks about having a baby and providing for it.

If we want to help someone like Monica work this through, we will need to do far more than pay lip service to her concern about the physical climate. We need the empathy that comes from having taken the issue on board ourselves. To remain sensitive to people – especially the young, who will bear the brunt of the climate crisis – we must remain vigilant in assessing whether our clinical setting has become a psychic retreat from climate reality. We cannot spare our patients the pain of facing the climate emergency, but we might be better able to help them if we ourselves understand the felt impacts of that emergency. We can only do that by undertaking the difficult work of becoming more climate aware ourselves. As psychoanalysts, we have much working through to do as we emerge from the climate bubble. We need the support of our analytic community and a culture of care to perform this work together. It seems to me that if we are able and willing to do the working through now required of us, this in itself may offer some hope to a patient like Monica.

Notes

1 First presented at a Scientific Meeting of the British Psychoanalytical Society on 17 November 2021.
2 E. Marks and C. Hickman, "Young People's Voices on Climate Anxiety, Government Betrayal and Moral Injury: A Global Phenomenon", *The Lancet*, 7 September 2021, https://papers.ssrn.com/sol3/papers.cfm?abstract_id=3918955 [accessed 30 July 2024].
3 It of course does not follow that all members of this cohort will not have children; also, other factors such as financial and housing prospects may influence their decisions.
4 Above preindustrial levels.
5 The *Oxford English Dictionary* reported on its website that according to the Oxford Monitor Corpus of English, *climate emergency* was used 76 times more frequently in the first half of 2021 than it was in the first half of 2018, and *climate crisis* had increased nearly twenty-fold over the same period. *Climate emergency* saw a spike in 2019, and since the second half of 2019 *climate crisis* has overtaken it in frequency.
6 This has in large measure been – rightly in my view – due to efforts not to pathologize people experiencing such feelings or to see them as ripe candidates for pharmacological medication.
7 See Sally Weintrobe, "Moral Injury, the Culture of Uncare and the Climate Bubble", *Journal of Social Work Practice*, 34.4 (2020), pp. 351–62.
8 See C. F. Alford, "Depoliticizing Moral Injury", *Journal of Psycho-Social Studies*, 9.1 (2016), pp. 7–19.
9 This is a specific application of John Steiner's concept of the psychic retreat, applied to the group and taking into account ideas on collective bubbles and financial bubbles. See John Steiner, *Psychic Retreats* (Routledge, 1993).

10 Sally Weintrobe, *Psychological Roots of the Climate Crisis: Neoliberal Exceptionalism and the Culture of Uncare* (Bloomsbury, 2021).

11 Also unbearable because of a general lack of cultural support to manage these feelings.

12 See Sigmund Freud, *Formulation on the Two Principles of Mental Functioning*, Standard Edition XII (Hogarth Press, 1911), pp. 213–26.

13 Nick Breeze, "ClimateGenn (2021) Inside #COP26Glasgow; Is This the Ambition Needed to Avert Disaster? Discussion Prof. Kevin Anderson", https://www.youtube.com/watch?v=5GcSxVHAxW0&t=18s [accessed 30 July 2024].

14 This position is beautifully argued and fleshed out with vivid examples in a paper on the setting by Jane Temperley, "Settings for Psychotherapy", *British Journal of Psychotherapy*, 1.2 (1984), pp. 101–11.

15 James Strachey, "The Nature of the Therapeutic Action of Psychoanalysis", *International Journal of Psychoanalysis*, 50 (1969), pp. 275–92.

16 See, for example, Paul Hoggett, "Climate Change in a Perverse Culture", in *Engaging with Climate Change: Psychoanalytic and Interdisciplinary Perspectives*, ed. by Sally Weintrobe (Routledge, 2012).

17 See Weintrobe, *Psychological Roots of the Climate Crisis*, parts Four, Five and Six, pp. 103–77. Here I argue that the dominant current culture is a culture of uncare, meaning a culture designed to boost uncare and undermine care.

18 See Weintrobe, *Psychological Roots of the Climate Crisis*, pp. 177–83.

5 Reflections on Plastic in the Sea and Other Transformations

A Significant Dream[1]

Alfredo Lombardozzi

A View of the Environmental Crisis: A Complex and Difficult Panorama

In the contemporary world, a feeling of unease is becoming more pressing, both on the individual and collective levels, because of the ever more serious and evident situation linked to the increase in global heating and the climatic and environmental crisis. The unease takes different forms on different levels, and I believe it is necessary to set the historical and social context in which we are living so as to make a more focused and detailed study of the potential for psychoanalysis to shed light on the unconscious dynamics which come into play in the attempt to deny the reality of climate change and to protect us from the anxieties it provokes.

Recently, with the coronavirus emergency and its consequences for personal and social relationships, we have undergone an experience which has made clear how necessary it is to make use of psychoanalytic tools to address a powerful social anxiety, when some apparently consolidated and shared resources are diminished.

Such a complex topic cannot fail to concern psychoanalysts since it involves the quality of the environment in which we live and its impact on the processes which generate it and have a close relationship with our psychic well-being – both at the individual level and in its social components, which are not confined to the narrow, local scale but affect the destiny of the whole human race. I would like to begin with some personal observations. However, when we talk about climate change, as we do today, especially in terms of the COVID-19 emergency, subjective events correlate strongly with the collective, social, and environmental components.

In a recent interview given during a conference on the relationship between psychoanalysis and the environment, I recalled the spring of 1986, when we learned through the media about the tragedy at Chernobyl. I remember my anxious thoughts as I watched my son, then a year old, playing in a garden, and wondered about the kind of future that our generation of young fathers was preparing for the children of the new generations.

DOI: 10.4324/9781003634959-7

I felt a sense of responsibility and, at the same time, powerlessness. I also remember the earnestness of our belief that we should be careful about the kind of milk or supposedly "uncontaminated" food we could provide for him. Indeed, after a few days, it became hard to find long-life milk, which we believed would be safer. It was like living in anticipation of a simultaneously pre- and/or post-cataclysmic (or pre- and/or post-traumatic) experience, resulting from a deep wound caused by an injury inflicted on the earth we live on (Weintrobe 2019).[2]

In those years, ecological thinking took Gregory Bateson's work and ideas as its point of reference, combining various observational viewpoints – psychological, anthropological, ecological, and cybernetic – into a new "ecology of the mind". Bateson proposed a new way of thinking about the problem of survival. To limit the survival of an individual, group, or living or cultural system solely to its existence – defined in strictly and traditionally Darwinian terms, detached from the dynamics of the system – would mean the destruction of the environment and of the very "subject" who had been responsible for the damage. According to Bateson, it was necessary to go in another direction – one that entailed a synergy between individual and environment.

From this perspective, survival would be determined by a close relationship between organism and environment, and it would be desirable to find a new way of thinking – and perhaps also of "acting" – which might embrace the view defined as *organism plus environment*.[3] As Bateson himself maintained, Freud had extended the mind – or that aspect of it that is unconscious – towards the interior, and we needed the "new way of thinking" to enable us to extend the mind towards the exterior, in both its conscious and unconscious components.

Bateson foresaw that if governments, scientists, and politicians – maybe the human race as a whole – did not acquire a "new way of thinking", then within 30 years the environmental situation would assume larger and more worrying critical dimensions. His forecasts have proved absolutely correct, and this is why it is important for us to look more closely at the reflections which psychoanalytic thought has offered – and continues to offer – about the relationship between humankind and the environment. The mechanisms of negation and disavowal, as well as the most openly denialist policies and positions, induce us to examine the topic from the viewpoint of the relationship between the internal world and external reality in more flexible terms. It is a question of finding another field of analysis which draws on what is today called the "extension of the psychoanalytic method".[4]

Returning to a brief historical survey, in the decades since the Eighties, despite the advance of a "green culture" and the important event of the 'Earth Summit' in Rio, the scale of the environmental crisis has become more and more alarming, with ever more obvious signs of an acceleration connected to the development of the "negative" aspects of globalization.

These have not only accentuated economic and social inequality but have also increased the exploitation of fossil fuels, causing a rise in the percentage of CO_2 in the atmosphere. Furthermore, there has been a weakening in those "positive" aspects of globalization which would – and to some extent do – enable us to find shared and negotiated solutions in terms of economic sustainability and cultural values.

In this context, Vittorio Lanternari, a teacher of ethnology and the history of religion, published in 2003, at the age of 85, an important book entitled *Ecoanthropology* – a book that I loved both for the friendship it brought me with Lanternari himself and for its far-sightedness and broad range in addressing the topic of the anthropological and ecological emergency.[5] The book explores the theme of the environment from an anthropological perspective, touching on many subjects: from the rapport which is established with nature in so-called "primitive" societies by the self-limited use of resources, to the neo-pagan religious movements originating in feminism in which the rapport with nature is central; the role of globalization processes; the exploitation and "inhumanity" involved in factory farming; the problems of diminishing biodiversity caused by deforestation; the spread of monocultures; and so on.

However, I would like to dwell on the central concept in Lanternari's thinking because I also consider it to be of great interest as a framework for thinking psychoanalytically about the environmental crisis. Lanternari is the spokesman for a critique of the position of Western man, which he calls *dogmatic anthropocentrism,* and which constitutes the manifestation of the tendency to consider human needs – especially those connected to economic profit and consumption – as being at the centre of existence and survival processes, according to the criteria of a disrespectful relationship with "nature", considered as an "otherness" to be exploited. Set against this principle is *ecocentrism,* a conception of the world which we could call "spiritual" – a deep ecology which puts "nature" at the centre rather than human action. Lanternari invites us to move beyond this dichotomy and look for a synthesis and a new form of unity in defining a field which may permit a relationship of productive exchange between what we could call natural forms of existence and forms of humanity – a dimension that he calls *eco-anthropo-centrism.* In this sense, though using a different vocabulary, he shares Bateson's complex vision of the relationship between man and nature as a system that is open and unified at the same time. The problem posed by Lanternari is highly topical in that, addressing what is presented today as a climate crisis in terms of an emergency with potentially – and ever more concretely – catastrophic outcomes, it posits the need to look for constructive forms of mediation in order to find sustainable solutions which take account of the "human" – or, in the broadest sense, anthropological – condition, which includes economic, cultural, and psychological factors, and at the same time allows us to erect a bulwark against the acceleration of climatic and environmental factors and the disequilibrium of ecological systems.

At the same time, what Lanternari proposed in coining the term *eco-anthropo-centrism* allows us to think that the very space – both conceptual and experiential – where the linking of the two terms, *ecological* and *anthropological*, happens may, in its many permutations, be the "psychic place" par excellence where human and non-human factors relate to each other in multiple ways, harmoniously or in conflict, also giving rise to fractures and possible consequent reparations.

In its broadest sense, the eco-anthropological dimension is the frame which, in my opinion, enables us to foster the dialogue between different forms of knowledge. At the same time, it is important to deepen the analysis of the psychic processes tested and studied by psychoanalysis in its clinical practice and experience, and which, when appropriately linked to the ecological dimension without arbitrary reductions or simplifications, may equip us with further tools for understanding the feelings and anxieties which the grave situation caused by the environmental crisis is producing and exacerbating.

I think that working within a framework that is both delimiting and flexible is fundamental if we are to cope with the various reactions which the ecological and environmental emergency generates on an emotional level, and in the theoretical responses and solutions we offer in relation to social, economic and environmental policies. For example, it is a very different matter to assess the use and role of technology and the economic forms correlated with it, depending on whether one adopts radically critical positions which see technology as a reification of the neoliberal "mentality", or, like Jeremy Rifkin, hypothesize a generative development working via online relationships and through self-organizing groups on the internet in order to optimize both communication and production processes, reducing their environmental impact to a minimum, if not to zero.[6] Then again, there is also a well-known difference between those who operate within the perspective of sustainable development and those on the side of degrowth.[7]

The anthropologist Thomas Hylland Eriksen speculates that there is a type of relationship between the two terms *development* and *sustainability* which can be likened to what, on the plane of psychic relations, Bateson has called a *double bind*.[8] This is something more than a dilemma – rather, it is a revelation of the difficulty in making choices which take account of strongly contrasting demands and needs. As a result, we are struck by the enormous complexity we are faced with, and baffled when we pause long enough to reflect on it.

We need only think of the scientist James Lovelock, father of ecology and the holistic Gaia theory, who, in the early 2000s, wrote a much-criticized book which, addressing the greenhouse effect caused by, among other things, the unrestricted use of fossil fuels, indicated as a possible (least bad) solution the encouragement of nuclear energy, with all the risks it entailed (Lovelock 2006).[9]

In his new book, *Novacene*, the same Lovelock, now aged 100, suggests an entirely new perspective. Reconsidering his original idea of Gaia as a system with a homeostatic capacity – an expression of substantial flexibility in environmental responses – he rather daringly hypothesizes a future denoted by the development of a new species of cyborg, parallel to *Homo sapiens*, characterized by increased intelligence and adaptability, and presumably inclined towards protecting – and even improving – the integrity of environmental conditions, in a vision that extends towards a significant connection with the functioning of the laws which govern the universe.[10]

Beyond the solutions that can be thought up, and which various "subjects" in the spheres of science, economics, or politics can propose, we are nevertheless faced with a reality which gives us the clear sensation of "running out of control":

> One way of describing overheating is to see it as the confluence of several *runaway* processes, forms of growth that were meaningful and purposive for a long time before reaching a point where the unintended side-effects were threatening to become more noticeable than the intentional or functional effects.[11]

It is no accident that, in 2000, the Nobel laureate Paul Crutzen adopted the term "Anthropocene", coined in the 1980s by the biologist Eugene F. Stoermer, to define the present geological epoch.[12] As we know, scientists today have accepted this suggestion, although the Anthropocene has not yet been formally recognized as a true geological era distinct from the Holocene. The Anthropocene is by definition a geological epoch, insofar as the activity of *Homo sapiens* has progressively, from the Industrial Revolution onwards, had such a significant impact on the very structure and ecological equilibria of the earth as to radically transform the climate and configuration of the globe, as is happening now.[13]

The serious problem, as Bateson had already anticipated and as Telmo Pievani suggests in his recent book *La terra senza di noi* [The world without us], is not only the accelerated extinction of animal and plant species on a scale that has never occurred before except at particularly critical moments in the evolution of life on earth, but rather that it is the human race itself which is endangering its own survival with its blind and denialist behaviour, creating the conditions for its own extinction.[14]

Just as our species *Homo sapiens* is, from this point of view, the result of contingencies which happened to occur, favouring it over other human species which became extinct, the conditions for the end of its existence on earth can be created in the same way. But the earth, or "nature", can rely

on the functioning of its homeostatic equilibria in order to continue existing despite its turbulences – at least until the implosion or explosion of the
sun and the entire solar system, which I believe is estimated to happen in
5 billion years.

Times and spaces on a scale to make one shiver, but Pievani urges us
not to be overwhelmed by catastrophic experiences, since – precisely
because *Homo sapiens* is strong and effective enough to produce such an
impact on nature – we can also hope that the necessary conditions and
tools are available to help us face the ecological crisis, obviously within
the limits imposed by the gravity of the crisis and the fact that we have, in
some respects, reached "points of no return". While not giving in to a pas-
sively catastrophist view and suggesting the possibility of following proce-
dures and lifestyles which may concretely and effectively slow the effect of
human activity in determining the environmental crisis, the meteorologist
Luca Mercalli warns us that *"time is up"*.[15]

Ecology and Psychic Processes: Extensions
of the Psychoanalytic Field

Returning to the main topic, it seems evident that psychoanalysis can make
a very important contribution to such a crucial subject in contemporary
life as the environmental crisis. A book published in 2020 by Cosimo Schinaia offers an expansive survey of how psychoanalysis has contributed
to reflection on the relationship between humans and the environment.[16]
Schinaia highlights the complexity of Freud's thinking and his attitude to
nature. In fact, if we go back to Freud's observations in *Civilization and its
Discontents* (1929) on the relationship between man and nature, his vision
emerges as one linked to the spirit of his time and directed towards an
emphasis on humanity's striving, through the evolution of technology,
to subjugate the uncontrollable and destructive aspects of nature, while
believing that progress in this direction would nevertheless not be a reason
for happiness for the human race.

On the other hand, in the brief but exquisite paper *On Transience* (Freud
1915) and his more personal reflections in the letters, Freud displays a
strongly aesthetic sensibility towards nature and an environment which,
in synergy with the sense of human precariousness and the feelings of loss
and mourning, becomes a sort of travelling companion in the search for
a more reflective and perhaps consoling intimacy. Nature consequently
exposes us to a powerful experience of strangeness and, at the same time,
to feelings of participation in a beauty which benevolently engages our
senses in meaningful moments of integration.

However, the scale of the present problem, which has been created over
time, has reached such broad and global levels as to entail highly complex ways of functioning on the mental plane. As I noted at the start, Sally

Weintrobe, who has edited a valuable book containing contributions by psychoanalysts and social scientists, has identified three ways in which defence mechanisms respond to this strong feeling of disorientation about the size and complexity of the problem caused by the climate and environmental crisis, and the anxieties to which we are all exposed.[17] The feelings behind the rejection of problems linked to the environmental crisis and climate change are categorized into three types. *Denialism* involves an intentional distortion of reality with the aim of implementing economic policies designed to oppose the choice of new technologies which favour the use of renewable energies for a more sustainable development on the environmental plane. *Negation* involves a tendency to defend oneself against anxieties of loss and mourning, but without distorting reality, and in this respect keeps open the possibility of regaining contact with problems that are hard to tolerate on the psychic plane. *Disavowal*, which, as we know, entails a greater defensive rigidity, involves taking refuge in an alternative reality in an attempt to rid oneself of distressing content as a source of disturbance. This means shifting attention from problems which are strongly conditioned by the impact of environmental changes we are living with, treating them – and psychically experiencing them – as if they did not exist.

The reasoning proposed by Weintrobe and reprised in various forms by the other authors of the text she edits is significantly widened by Joseph Dodds who sheds light on the complexity of the theme he is exploring, taking as his starting point an interdisciplinary approach and making use in particular of non-linear systems and the thinking of Deleuze and Guattari. On the psychoanalytic plane, as Schinaia aptly describes, Dodds refers to Kleinian positions, "including the phantasy of an infinitely available earth-breast, the paranoid-schizoid response to the necessity of weaning, and the need to move towards a depressive position, with its associated desire for reparation for loss, pain, and disappointment".[18]

I will now dwell for a moment on the contribution of Searles who, back in the seventies, in an essay of great interest, had anticipated some of the themes relating to the primary anxieties and their relationship with the ecological crisis. Picking up the theme of the Oedipal conflict in reverse, Searles maintains that the anxieties about environmental pollution, with its damaging effects on the future of the human race, are expressed in terms of *apathy* and *indifference*, for the very reason that generations of adults, or the elderly, tend unconsciously to adopt a rivalrous stance towards younger generations – to such an extent that they become unconcerned by the fact that the earth or the human race may be destroyed or go into decline. Viewed in this way, becoming old would not entail the achievement of wisdom so much as a form of "moral pollution" projected onto the environment, resulting in the feeling "that the pure air and water and so on of our childhood is now lost forever".[19]

In this way, as he sees it, we share a strongly depressive feeling in considering that the earth and nature, which we have wounded, are in inexorable decline and that the pure world of childhood is no longer recoverable. In this emotional scenario, we feel impotent in being unable to do anything for the earth and, at the same time, omnipotent because "we" have been the cause of the wound that is degrading it. Searles believes that the greatest psychological risk in the environmental crisis is the apathetic attitude determined by the inability to address the primary conflict between the human and non-human parts of the Self, the latter being projected outwards and consequently identified with an omnipotent, controlling, destructive technology that depletes natural resources. In the final analysis, unable to tolerate the conflict between human and non-human in the Self, "we project this conflict upon, and thus unconsciously foster, the war in external reality between the beleaguered remnants of ecologically balanced nature and man's technology which is ravaging them".[20] From a social anthropological viewpoint, Mauro Van Aken recovers the notion of the uncanny in the Freudian sense, which nevertheless, in the processes of modernity affecting the rapport between man and nature, becomes a cultural rather than a psychic operation – an "othering" of nature that causes it to be considered a foreign entity to be subjugated, activating the forms of negation and disavowal to which we have already referred. In his fieldwork on what we could call "Water culture" in Jordan and other Middle Eastern contexts, Van Aken highlights the importance which non-Western cultures attribute to the ability to enter into a relationship with water as a natural resource, in a process of integration between cultural aspects and natural elements, between forms of humanity and the non-human environment.[21] So it seems very important to me that we should be able to grasp the drama of the relationship with nature when it is understood as an environment which includes us, while also manifesting and generating an uncanny sense of strangeness, characterized by the complexity of the problem and the scarcely conceivable scale of the risks entailed by climate change.

However, I think it is also important that we integrate these considerations with some reflections which draw on two features which Heinz Kohut has illuminated throughout his work. In his essay "Introspection, empathy, and the semicircle of mental health", on the theme of the relationship between the generations, Kohut offers a very interesting interpretative key. He reports the mythical episode which recounts the moment when the Greek emissaries visit Odysseus to persuade him to take part in the Trojan War. Odysseus, who had no intention of taking part in the war, pretends to be mad, ploughing a field with oxen and sowing it with salt.

Palamedes detects the deception, picks up the baby Telemachus, and throws him in front of the oxen, forcing Odysseus to make a semi-circle and save his son. Kohut offers this story as a metaphor, stressing that

> the semi-circle of Odysseus' plough [. . .] is a fitting symbol of that joyful awareness of the human self of being temporal, of having an unrolling destiny: a preparatory beginning, a flourishing middle, and

a retrospective end; a fitting symbol of the fact that healthy man experiences, and with deepest joy, the next generation as an extension of his own self.[22]

Moving onto a macrosocial dimension, I find it helpful to return to Kohut's concept of the "cultural selfobject", broadening it – rather as Bateson does – into a meaning closer to that of "organism plus environment", with the suggestion that it be reformulated as the "cultural/environmental selfobject". The cultural self-object is that psychic figure which enables human groups, and the cultures that define them, to recognize themselves in a form of identity construction of the group-Self, in such a way as to acquire a "relative" sense of cohesion and continuity through the sharing of a common ideal.[23] In these terms, what I propose to call the "cultural/environmental selfobject" may attain a relationality with the "natural" forms, not perceiving them only as "forms of strangeness", but also in the dynamic articulation of the profound intertwining of nature and culture.

The inevitable contingencies to which we are all exposed, the unpredictability of the movements of natural forces, and the equally inevitable frustrations and conflicts in interhuman relations and in those between cultures create deep fractures and breaks which, as we have seen, are today compromising the very equilibrium of eco-anthropological systems in the form of excess and acceleration. It is possible that our "future" depends on the extent to which – if we wish to use a psychoanalytic terminology – feelings and actions aimed at making *reparative choices* can counter the grave threat to the our striving for unity with the environment.[24]

Reflections on Plastic in the Sea and Other Transformations: A Significant Dream

I would now like to try to link the macrosocial themes connected to the climate crisis and environmental questions with some clinical evidence. For this purpose, I think it will be helpful to draw on a highly vivid and complex dream of a young adult patient whom I will call F, "the man from the sea". The dream is told in three parts. In the first part, he finds himself beside a rough, stormy sea in a gloomy twilight and sees a lot of white plastic knives floating on the surface of the water, creating something like a shining island.

The image is very vivid and, in some respects, refers to an experience connected to a bereavement – the loss of his mother some years earlier. The marine setting in particular and certain associations suggest a maternal relationship with symbiotic features and the revelation of a traumatic break which has produced painful experiences in him, together with a strong urge towards autonomy: an urge which had manifested itself more obviously when he was young and is still expressed today in his careful management of the small house he lives in. This compels him to maintain a difficult balance between aspects of himself that are embodied in significant objects he constructs by hand from everyday materials and "recyclable" or "non-biodegradable" waste (he had recently dreamed that he was throwing rubbish into bins). The experience of mourning had exposed him to experiences of rage and abandonment, bringing him into contact with a rejected Self. In the dream, he enacts an experience related to split-off aspects of the Self which, in terms of the ecological metaphor, take the form of plastic knives that are non-degradable or, on the level of intrapsychic experience, cannot be worked through or integrated.

In a second part of the dream, F is in a museum. The light is still dim. In one of the rooms, the bust of a strange figure is on display: a man looking like a pirate, with parts of his body resembling the tentacles of an octopus or squid. We share the association with the character of Davy Jones, a pirate in the *Pirates of the Caribbean* series, who has these same features: chimerical figures mixing human elements with those of aquatic creatures and various types of cephalopod and crustacean. The image seems to portray an isolated aspect of the Self, enclosed in the "museified" form of a statue into which conflicting human and non-human parts of the Self – as Searles puts it – are compressed.

In the third part of the dream, F again finds himself in a marine setting, but a very different one. It is sunset and the sun is warm – a hot, enveloping pink. The sea is calm and barely moving. F walks along the shore, passing near and slightly above a row of those boulders which are banked up into dykes to stop the sea rising above a certain level. He is thinking pleasant thoughts and feels tranquil. The sea is an element of his homeland, and it seems that this third image has something to do with a form of pacification that can occur within the dream as a whole – which simultaneously represents different levels of intrapsychic experience – where a feeling of reparation takes shape both in relation to feelings of loss, the original wound, and the experience of abandonment, and to the destructive fantasies about the relationship with a mother who is perceived as destroyed or absent and potentially vengeful.

These dreams seem to express several permutations of denial since the images convey something grand, like the "shoal of shining plastic knives", which has its own beauty but is also "polluting", and, at the same time, the troubling image of the bust of the aquatic-chimerical man – powerful but enclosed and compressed in a claustrophobic state: a dream which, while it dramatizes the reasons for a grand denial, also highlights the risks linked to the anxiety that the "wounded" mother – who is also "mother nature" – may carry out harsh retaliations, setting terrifying non-human forces in motion.

Now, it also needs to be said that environmental themes were not especially present in F's analysis on a conscious level, but if we widen our perspective, we notice that the dream's images express the power of an unconscious possessing features of both nature and culture. We also detect a highly complex dynamic through which the estranging non-human, animal, inorganic, and artificial forms coexist with reparative tendencies that can be likened to the stabilizing functions of a self-object which "culturizes" the relationship with the environment – sea/maternal – erecting a dyke against otherwise unmanageable emotions in a rediscovered emotional warmth.

In the dream, though without an explicit reference to an environmentalist commitment, F's intrapsychic world portrays a powerful and implicit relationship of complementarity with the macrosocial dimension, which relates to anxieties about the environmental crisis and attempts to find ways of confronting it. The fluidity of the images of the "oneiric unconscious" permits the identification and introduction of feelings on the individual and personal level, which may be able to challenge that experience which, on a more "general" and complex level, has been called "The Great Derangement". This may enable us to achieve a greater awareness of the scale of the climate crisis and encourage us to take "responsibility" for our affects, both on a personal level and in relation to the state of human life on earth. The "aesthetic" and "metaphorical" context of the dream's third part, in the "relative" equilibrium between marine and atmospheric agents, makes us perceive the extent to which tides and turbulence are constantly moving "inside" and "outside" us, since we are always looking at the sky and being "weathered" by it, as Van Aken tells us in his recent book.

Challenging the "Great Derangement": Towards a Psychoanalytic Anthropology of the Future

The recently formed *Fridays for Future* movement and the increased use of renewable energy sources give us hope, but this is all insufficient if conditions are not created for demolishing denialist policies and if an effort is not made – even on a small scale – to shed light on the workings of negation and denial, to which we all contribute in our own way. These processes create splits which allow us to see and not see, preventing us from fully appreciating the severity of the predictably catastrophic developments now manifesting in changes to the climate. As proclaimed by the great majority of scientists and by the youth movement through the testimony of Greta Thunberg, "time is short": to a considerable extent, the damage has already been done, and those defence mechanisms which the writer Amitav Ghosh has effectively summed up as the "Great Derangement" are starting not to work.[25]

In his recent reflections on the environmental crisis as an "unthinkable" situation, Ghosh makes us take a serious look at the processes of negation at work not only in individuals but also in the policies which have effectively removed the collective dimension of the relationship between humans and the environment. We should instead acknowledge the fact that, in the relationship between the human and non-human environments, the boundaries do not correspond to the artificial ones of nationality. As Ghosh suggests, the collective dimension which is inherent in the relationship between the human and non-human environments is almost completely neglected in the contemporary "bourgeois" novel, which exalts an individual psychology of affects rather than talking about our immersion in a complex world where the human and non-human coexist.

As I have maintained during these reflections of mine, this results in the setting up of unconscious dynamics to deny the "uncanny" factors of the non-human, which are thereby presented in exaggerated forms and on such an "unthinkable" scale that they cannot be addressed. Ghosh maintains that it is as important to bring about an evolution in the practices and presuppositions which guide the arts and humanities as it is to develop a complex technical language for climatic changes: "Indeed, this is perhaps the most important question ever to confront culture in the broadest sense – for let us make no mistake: the climate crisis is also a crisis of culture, and thus of the imagination".[26]

Consequently, from a psychoanalytic viewpoint, as Luc Magnenat maintains, this is the moment for turning "catastrophe into dream" – in a certain sense, "to think the unthought" – and to rediscover an Ethical dimension founded on the awareness that the relationship between man and nature (human–non-human) is shot through with unconscious dynamics and predominantly Oedipal drives which, if well managed on a collective level, may enable us to build a more synchronic relationship between culture and environment.[27]

It seems to me that, even if we draw on very different theoretical models, a psychoanalytic reading of the critical state of the relationship between man and environment may help us bring complex forms of relationship and unconscious dynamics into being, in such a way as to challenge the tendency towards various kinds of negation, encouraging the search for reparative modes that will facilitate a new relationship founded neither on a dichotomy nor on an excessive and improper assimilation between nature and culture.

So, we can imagine and think about a new Ethic which allows us to put forms of negotiation between different values into effect in the search for shared ways of guaranteeing that new generations, even within a relatively short timescale, can be buttressed against climate change and the disruptive effects that are manifesting themselves today.

This means fostering the striving to aspire to what Lanternari has called "Added Value": that is, an emblem of the ethic that will guarantee the future of the generations to come, inspired by the *Imperative of Responsibility* of Hans Jonas – one that may allow us to look for common ground in order to address our ecological problems, creating a dialogue of differences without cancelling them out or turning them into inequalities, as is often the case.[28] Perhaps this process would facilitate the sharing of the feeling that we are "all in the same boat", and be a way to challenge the sense of a world ending with no redemption, as Ernesto De Martino has taught us in his notes on *The End of the World*, aptly describing how the end of "nature" is also – and especially – the end of any possible form of culture.[29]

The set of considerations that I have proposed emphasizes how desirable it is to have an ever-closer dialogue between the humanities, the eco-biological sciences, and psychoanalysis in sharing a passion for a common knowledge that may enable us to know the environment we live in, so that we can intervene in it more appropriately. The contribution of psychoanalysis is fundamental in creating a field of reflection which may serve as a bridge between clinical experience, the clinical models we work within, and the social forms of suffering in the relationship between humans and the environment.

Notes

1 This chapter originally appeared as "Climate Change and Environmental Crisis: Psychoanalytic Thoughts Towards an Anthropological Ecology", *The Italian Psychoanalytic Journal* (2021), pp. 57–71. Translated by Adam Elgar.
2 See Sally Weintrobe, "Vivere con i nostri sentimenti sul crollo dei sistemi planetari", paper read at the conference *Cambiamenti climatici. Crisi ambientale, angosce e forme della negazione*, Centro di Psicoanalisi Romano, Rome, 9 November 2019.
3 Gregory Bateson, *Steps to an Ecology of Mind* (The University of Chicago Press, 1972).
4 T. Bastianini and A. Ferruta, eds, *La cura psicoanalitica contemporanea: Estensioni della pratica clinica* (Fioriti, 2018).
5 Vittorio Lanternari, *Ecoantropologia. Dall'ingerenza ecologica alla svolta etico-culturale* (Dedalo, 2003).
6 Jeremy Rifkin, *The Green New Deal: Why the Fossil Fuel Civilization Will Collapse by 2028 and the Bold Economic Plan to Save Life on Earth* (St Martin's Griffin, 2019).
7 See V. Termini, *Il mondo rinnovabile. Come l'energia pulita può cambiare l'economia, la politica e la società* (LUISS University, 2018), and S. Latouche, *Le pari de la décroissance* (Fayard, 2006).
8 Thomas Hylliand Eriksen, *Overheating: An Anthropology of Accelerated Change* (Pluto Press, 2016).
9 See James Lovelock, *Gaia: A New Look at Life on Earth* (Oxford University Press, 1979), and *The Revenge of Gaia* (Basic Books, 2006).
10 James Lovelock, *Novacene: The Coming Age of Hyperintelligence* (MIT Press, 2019).
11 Eriksen, *Overheating*, p. 22 (My italics).
12 Paul J. Crutzen, *Benvenuti nell'antropocene*, ed. by Andrea Parlangeli (Mondadori, 2005).
13 Simon Lewis and Mark A. Maslin, *The Human Planet: How We Created the Anthropocene* (Pelican Books, 2018).
14 Telmo Pievani, *La terra dopo di noi, Fotografie di Frans Lanting* (Roberto Koch, 2019).
15 L. Mercalli, *Non c'è più tempo. Come reagire agli allarmi ambientali* (Einaudi, 2018).

16 Cosimo Schinaia, *L'inconscio e l'ambiente: Psicoanalisi ed ecologia* (AlpesItalia, 2020).
17 See Sally Weintrobe, ed., *Engaging with Climate Change, Psychoanalytic and Interdisciplinary Perspectives* (Routledge, 2013).
18 Schinaia, *L'inconscio e l'ambiente*, p. 54. See also Cosimo Schinaia, "Respect for the Environment: Psychoanalytic Reflections on the Ecological Crisis", *The International Journal of Psychoanalysis*, 100.2 (2019), pp. 272–86 (p. 54).
19 See Chapter 7, Harold Searles, "Unconscious Processes in Relation to the Environmental Crisis", p. 84.
20 Searles, p. 87.
21 Mauro Van Aken, "Vivibilità e crisi ambientale, dismisura e vivere sopra", in *Vivere Sopravvivere*, ed. by A. Lombardozzi (AlpesItalia, 2018); and Mauro Van Aken, *Campati per aria* (Eléuthera, 2020).
22 Heinz Kohut, "Introspection, Empathy and the Semi-Circle of Mental Health", *The International Journal of Psychoanalysis*, 63 (1982), pp. 395–407 (p. 404).
23 Heinz Kohut, "On the Continuity of the Self and Cultural Selfobjects", in *Self Psychology and the Humanities*, ed. by C. Strozier (Norton, 1981).
24 Alfredo Lombardozzi, "Gaia. Riflessioni a proposito del libro Ecoantropologia di Vittorio Lanternari", in *Figure del dialogo tra antropologia e psicoanalisi* (Borla, 2006).
25 Amitav Ghosh, *The Great Derangement: Climate Change and the Unthinkable* (The University of Chicago Press, 2016).
26 Ghosh, p. 9.
27 Luc Magnenat, "Penser comme une montagne, penser oedipien: une contribution psychanalytique à l'éthique environnementale", paper read at the conference *Cambiamenti climatici. Crisi ambientale, angosce e forme della negazione*, Centro di Psicoanalisi Romano, Rome, 9 November 2019.
28 See Lanternari, *Ecoantropologia*, and Hans Jonas, *The Imperative of Responsibility* (The University of Chicago Press, 1984).
29 Ernesto De Martino, *La fine del mondo. Contributo all'analisi delle apocalissi culturali* (1977), ed. by Giordana Charuty, Daniel Fabre and Marcello Massenzio (Einaudi, 2019); and Alfredo Lombardozzi, "La fine del mondo. Attualità di Ernesto De Martino", *Vulnerabilità. Psiche*, 1 (2020) pp. 149–63.

6   Do Humans Really Want to Survive?

Don Moss

This dreadful question demands that we imagine the end of ourselves. For me, nothing softens the image of our end. I do not picture a prolonged decline; I see a frenzy of catastrophes transforming the planet, forcing every life form – humans, plants, animals – to become refugees, each seeking a sustainable environment. I see us humans, panicked by scarcity, unable to band together, fragmenting into competing clusters, suspicious and impoverished rogue units. There is not enough. We are too many. The image expands and explodes. The planet is swarming with atrocities, floods, disease – all the worst I've known. The forces of order have vanished. No law, no regulation. I think that perhaps I'm being excessive or self-indulgent. And yet, as though beckoned onto the page by my florid imagination, I read this in the news:

> Deforestation is now being driven by criminal networks that conduct sophisticated schemes to hide the illegal origins of commodities like beef, gold and timber from the Amazon and sell them on international markets. A 2022 study by the Igarapé Institute, a Brazilian think tank, found that though the "cluster of converging crimes" varies in each Amazonian region, it often involves violence, corruption, financial crimes and fraud.[1]

OK, read the stuff, but let it go, I think. Why keep on imagining this, why push towards this end? Why not instead imagine a close call, a near miss: a world of wind farms, solar energy, captured carbon, heroic scientists, generative activism, environmental justice, a new equilibrium – the Anthropocene's steep curve bottomed out, the human relation to the planet organized around stability, order and care?

Why not? Because only by confronting these catastrophic images and the spectre of the worst possible outcome can I explore the underlying question: do humans truly want to survive? This question only makes sense if the answer might be "No".

To think about the question psychoanalytically, I begin with the internal and external drives that make continuous demands upon our minds. The

DOI: 10.4324/9781003634959-8

work is complex, but the underlying demands are straightforward. From the inside: find food, find another, find sex, find a place, find safety. And from the outside: see us, recognize us, know us, locate us, obey us. All the elaborated possibilities for being found and protected, for love, for meaning, for something new, for a future, are grounded in this base cluster: find and protect your mind and body, find and protect your place and your people. These demands never end – we must always be working to either satisfy, eliminate or modify them. But is all this work actually worth it? In what follows, I take up this revised question: what might make the work of survival worthwhile, and what might make it not?

***

A patient repeatedly arrives at the following conclusion:

> I do not want to think. I do not want to be here. I don't want to speak. I don't want to see you. I don't want you to see me. I don't want to be a father, a husband, a son. Why am I doing any of this? I want it to end. I'm waiting for it to end. I want to be in the corner, where it's impossible to move. What's the point?

This patient never misses a session. He is never late. He calls me his life preserver, and yet repeatedly says he has no need for me. I am of no use. I will get sick of him. "It's going to end some time, so why not end it now?" He resents all demands on his time. Occasionally, he gets into a frenzy.

Nothing works; he can't escape. He throws things, bangs his head against walls, hits himself in the face, and calls me to say the treatment is over. "I'm out of control. I can't get on top."

He is "on top" only when he hooks up with "the guys" he meets on an online S&M site: "The guys are like objects, not human, sub-human, like animals, I can do anything I want, it doesn't matter, it doesn't count what I do to them."

"I will not be fooled," he says. "I will not be duped." He wants me to do all our work alone while he watches scornfully. "I will not work," he says, repeatedly. But in spite of this, he stays with it; he works. "It's very good to see you," he said just this morning, retracting it immediately with, "I want to stop." "I will not be a servant," he says, reminding us both of all the times he has "hooked up" with his sub-human objects.

The patient always has his eyes on me. Am I in this with him? Is this a con?

When he asks whether I am in this with him, I think what he wants – what he is very uncertain of getting – is a fair reward for whatever work he does. He is looking for what he has never received: a clear sense that he is not alone in working, that I too am at work, and that whatever rewards are available will be shared equitably between the two of us. He wants justice.

At its foundation, the work of psychoanalysis aims at just that: justice. By "aiming at justice", I mean we aim to give ourselves and our objects a fair opportunity to diminish the restrictive, "unjust" influence of transferred prejudgements.

The patient is a brown-skinned man born into a middle-class family in Southeast Asia. His country was a European colony for centuries, and its colonial vertical relations still persist without noticeable dilution. My patient therefore necessarily mapped himself and was mapped by verticality. His family had a permanent group of four or five servants who were paid almost nothing and who could be beaten whenever their work or their attitudes were judged to be lacking. They had no legal standing by which to resist. The patient never questioned the family's right to beat their servants.

Beating them was, in fact, encouraged – a sign that he was properly entering the social order.

The patient immigrated to the United States to "get away from my country" – its static social order – although he has never mentioned a wish to get away from his right to beat people. Without much explicit attention, though, the hook-ups with "the guys" have almost completely stopped. He intermittently targets and "beats" himself now, bloodying his head and fists by banging them against walls. The steep vertical set-up of master and servant has infiltrated and structured the treatment more subtly. "You are nothing," "you don't matter to me," "I have no obligation to you," he will frequently say – but, crucially, and perhaps like his servants, he will also say that I am his "life preserver".

Though it took some time to be established and named, the steep vertical set-up in the treatment room now appears to waver. "If only I could just say, 'it's good to see you' and know I meant it, but I can't. I can't tell if it's a trick or what it is. Something terrible must have happened to me. I can never tell what I mean. What happened?" When he says "it's good to see you", I experience a rush of the same feeling: it's good to see him. But the equality suggested by this shared encounter with horizontality is brief and unstable. The moment ends abruptly, as though a mistake has been made. Neither of us is permitted to feel good with or about the other.

He says, "That's what you want to hear, right, that it's good to see you? The truth is I just want to be left alone. Just leave me alone." I think he doesn't simply want to be left alone. I think instead he wants to be able to say "it's good to see you" or "it's bad to see you" – to say, in fact, whatever it is that he might mean, to have the right and the capacity to say it, and the confidence that not only will it be heard, but that I too will have the right to say what I mean, and that my words, like his, will be heard. The pursuit of this kind of reciprocity in psychoanalytic work is what I mean by "aiming for justice".

A brown-skinned man from a colonized country in Southeast Asia in dialogue with a white-skinned man from the imperial United States. Without explicitly naming that frame, each of us hopes to escape the prejudice of our vertically oriented pasts to allow for a straightforward – horizontal – exchange. The possibility of that kind of exchange, the shared pursuit of it, and the justice of it make our work worthwhile for both of us.

I think what is true for the two of us in our psychoanalytic work may pertain to human relations more generally: that the work of survival will only be worth it if we are in pursuit of justice for all. The possibility of justice – the grounding awareness that "I cannot do anything I want" – means sacrificing notions of dominion, ceasing to target and victimize objects, whether non-human, "sub-human", or "animal". This constitutes a kind of psychic decolonization in which "it's good, or even it's bad, to see you" replaces "it's good to have dominion over you". Such a movement towards the horizontal replaces preservation of the vertical.

Can we humans imagine and endure a planet-wide pursuit of justice towards each other and towards our fellow inhabitants of this planet? So far, I think we have not.

Two voices from Pakistan, documented in a 2022 *New York Review* article about unprecedented flooding, exemplify this:

"Obviously, the bargain made between the global north and global south is not working." [. . .]. "Is this life now? Will we keep getting drowned this way, again and again?"[2]

When voices fail, action follows. According to the same article, Mai Jori Jamali, a prospective parliamentary candidate for Jafarabad, demanded the opening of a water-control gate that, if kept closed, would lead to the flooding of local villages. When her plea failed, she announced that village children would refuse polio drops until this demand was met. The gate was then opened.

Sigmund Freud recognized in 1927 the concerns we are faced with now, though he posed a slightly different question. For him, it is not a matter of whether we want to survive, but whether we ought to:

It goes without saying that a civilization which leaves so large a number of its participants unsatisfied and drives them into revolt neither has nor deserves the prospect of a lasting existence.[3]

***

Humans have mapped the planet from the bottom of a vertical set-up. Think of the planet as a given – our ultimate object, the fundamental and immortal thing in regard to which, or through which, the drive is able to achieve its aim. Reduce the object to silence, as we have done to our planet, give it no say in setting the terms by which it can be put to use, and we find ourselves occupying the terrain of dominion – a relation of absolute authority, the position of master. The planet, its inhabitants, and all its organic surround are relegated to the position of slave. The biblical promise of Genesis seems upon us now, not as a gift, but as a curse:

> And God said, Let us make man in our image, after our likeness: and let them have dominion over the fish of the sea, and over the fowl of the air, and over the cattle, and over all the earth, and over every creeping thing that creepeth upon the earth. And God blessed them, and God said unto them, Be fruitful, and multiply, and replenish the earth, and subdue it: and have dominion over the fish of the sea, and over the fowl of the air, and over every living thing that moveth upon the earth.
>
> (Genesis 1.26)

The planet can offer no resistance, cannot rebel, cannot sabotage. It can only suffer, endure, and adapt to our efforts. We are in the midst of its current flurry of adaptations. The planet lacks consciousness. Viewed from the point of view of dominion, it is a speechless thing, inert except for its limited capacities to satisfy. In its relation to the planet, dominion is exerted through extraction – satisfaction its only goal.

Homer's *The Odyssey* begins when Telemachus sets off to find his father, Odysseus. But Telemachus requires approval before his departure. He needs wind and a calm sea, for each of which he needs not only permission but also cooperation and assistance from the deities Athena and Poseidon. He is not autonomous – he is a supplicant. The world is not his. He lives in a sacred space in which he needs cunning more than raw power. He needs to work with, not against, the gods. This relationship constitutes a stable triangle of the forces of regulation: humans desire, gods govern, and the planet, conditionally, provides.

Remove the gods, and the stable triangle vanishes to be replaced by an unstable dyad – humans and the planet – without a regulating authority. The frenzy of dominion can begin. In Joseph Conrad's *Heart of Darkness* (1899), Telemachus becomes Marlowe, Odysseus becomes Kurtz. Marlowe, our avatar, does the work of an imperial explorer, a man on a mission that begins in London and finishes in Africa. The planet and its peoples have been stripped of all meaning, except what can be found hidden in their depths and taken from them. As Marlowe mines the planet for what it can yield, the sacred terms of *The Odyssey* are transformed into the profane terms of *Heart of Darkness*. Now, 125 years later, the relatively discreet colonialism of *Heart of Darkness* has exploded.

Human beings regard the planet as a colony, with extraction as our primary aim. Find, take, satisfy. Find, take, satisfy. The refuse from our unending appetites has generated a swarm of climate refugees – human, animal, and plant. The unalloyed desire for dominion undermines the pursuit of justice and makes it impossible, and therefore also undermines the motivation to do the work necessary for survival. Dominion generates a gang mentality. The planet has become the gang's turf.

The psychoanalytic dyad is explicitly organized against the temptations offered by the promise of dominion. Neither the conscious nor unconscious fantasies of either party can claim legitimate authority. Unconscious fantasy infiltrates and, in effect, colonizes our minds. To work psychoanalytically is, in effect, to work anti-colonially – to name and undo the long-standing effects of distorting infiltrations, to liberate oneself. The promise of liberation makes our work worthwhile. I think the same holds for all of us humans. Without that promise, what is the point?

Epilogue

A group of climate activists in Iceland is working to nominate an iceberg for the nation's presidency. A journalist friend of mine, recently returned from writing about the group, tells me this story:

> We were interviewing on a nearly inaccessible lava mountain whose cliffs sloped down to the ocean. Before we began, we paid homage to the Huldufólk, or little people, whom, we learned, welcomed gifts of honey. We had a lot of it so we slathered some along the tops of a nearby semi-circle of rocks. We stared out to the ocean and immediately saw a seal, hardly moving, staring back at us. And then, directly beyond the sea, we saw two fin whales cresting. They went down for a while and came back up again. Now there were the six of us and the three of them. The three of them soon vanished. But just as they did, an arctic fox, Iceland's only native mammal, appeared at the edge of the semi-circle of rocks. He stared at us for a while and then began moving, slowly, from rock to rock, licking off all the honey.
>
> When he was finished, he left. It took a very long time before any of us could speak. I had been seen and welcomed as I had never before been seen and welcomed. For the first time in my life, I was home.

Home, I thought. That's it. Were all of us only home . . .

Notes

1 Heriberto Araujo, "For Lula and the World, the Tough Job of Saving the Amazon Begins", *New York Times*, 31 December 2022, https://www.nytimes.com/2022/12/31/opinion/brazil-elections-amazon-rainforest.html [accessed 25 July 2024].
2 Alizeh Kohari, "Pakistan Submerged", *The New York Review*, 24 November 2022, https://www.nybooks.com/articles/2022/11/24/pakistan-submerged-alizeh-kohari/?lp_txn_id=1567379 [accessed 25 July 2024].
3 Sigmund Freud, *The Future of an Illusion* [*Der Zukunft Einer Illusion*], trans. by James Strachey (1927; W. W. Norton, 1961), p. 12.

Part II
Mainly Theory

7 Unconscious Processes in Relation to the Environmental Crisis[1]

Harold Searles

Even beyond the threat of nuclear warfare, I think the ecological crisis is the greatest threat mankind has ever collectively faced. The stream of articles and books calling our attention to various aspects of this crisis comes from ecologists, population biologists, physicists, chemists, agriculturists, economists, architects, engineers, city planners, statesmen, historians and, mainly, concerned laymen – some of whom provide valuable insights into the psychological ingredients of the problem. But rarely, indeed, is a behavioural scientist heard from, and to the best of my knowledge, very few psychiatric articles have appeared as yet concerning this subject, and but one contribution from a psychoanalyst – Peter A. Martin of Detroit – who touched upon it briefly and incidentally in a talk I heard him give in April 1969.[2] This environmental crisis embraces – and with rapidly accelerating intensity threatens – our whole planet. If so staggering a problem is to be met, the efforts of scientists from all clearly relevant disciplines will surely be required. It seems to me that we psychoanalysts, with our interest in the unconscious processes which so powerfully influence man's behaviour, should provide our fellow men with some enlightenment in this common struggle.

My hypothesis is that man is hampered in his meeting of this environmental crisis by a severe and pervasive apathy which is based largely on feelings and attitudes of which he is unconscious. The lack of analytic literature on this subject suggests to me that we analysts are in the grip of this common apathy. A second factor – a particular felt hazard within our profession – tends to inhibit us from making the special contributions we could make: we fear that an active concern with this present subject will evoke, from our colleagues, nothing more than a diagnostic interest in whether we are suffering from psychotic depression or paranoid schizophrenia.

As to the evidence for the general apathy I postulate, our federal budget for 1971 includes only about one seventieth as much for dealing with environmental pollution as for military purposes.[3] I have no wish to speak lightly of our military needs, but it does seem evident to me that a citizenry actively aroused about the state of our – and the world's – ecology would not accept so feeble an effort in this area.

DOI: 10.4324/9781003634959-10

Of the mass of statistics concerning the environmental crisis, here are a few of the items that I find awesome. We are dumping into the ocean as many as half a million different pollutants, only a very few of which have been studied for their possible effects on ecologically vital processes.[4] There is increasing concern that in a few short decades the ocean may become incapable of supporting living creatures – as are already many of our great rivers, several of our Great Lakes, and the Baltic Sea.[5] The pesticide DDT, to mention but one pollutant, has been discovered as far afield as in the bodies of Arctic Eskimos, Antarctic penguins, and seals.[6] Within less than two decades of their introduction, "The synthetic pesticides have been so thoroughly distributed throughout the animate and inanimate world that they occur virtually everywhere."[7] Seventy percent or more of our planet's total oxygen production by photosynthesis occurs in the ocean and is largely produced by diatoms; recent studies have shown that DDT, which permeates all things in the ocean, impedes the diatoms' production of oxygen.[8]

As for the radioactive waste from atomic reactors, we are already heirs to some 80 million gallons, stored in tank farms. These tanks will have to be guarded for 600 to 1,000 years. Several storage tanks have already leaked thousands of gallons into the soil, and a single gallon is enough to poison a city's water supply.[9] David Lilienthal, the first chairman of the Atomic Energy Commission, has stated, "Once a bright hope shared by all mankind, including myself, the rash proliferation of atomic power plants has become one of the ugliest clouds hanging over America."[10]

The accelerating overpopulation of the earth is a factor of transcendental importance. It is estimated that in 6000 BC there were 5 million people on earth, and that it had required about 1 million years for the population to double from 212 to that 5 million. From then on, doubling occurred about every thousand years, reaching a total of 500 million (that is, half a billion) around 1650. Then the population doubled within only 200 years, reaching a billion by 1850. The next doubling took only 80 years, reaching 2 billion by 1930. The doubling time at present is about 37 years. If population growth were to continue at the present rate – which it obviously cannot – for another 900 years, there would then be about 100 persons for each square yard of the earth's surface, land, and sea.[11]

Famines, especially in underdeveloped countries with their higher growth rates, are one of the "solutions" to this clearly impossible situation. The population biologist Ehrlich notes that more than half the world is in misery now, and that an estimated 5 million Indian children, for example, die each year from malnutrition.[12] He is convinced that within this decade, hundreds of millions of people are going to starve to death in spite of any crash programmes embarked upon now.[13]

The United States, with less than one fifteenth of the world's population, uses well over half of all the raw materials consumed each year, and if present trends continue, in 20 years we will constitute much less than one fifteenth of the population, yet may consume around 80 percent of the world's resources.[14] On the other hand, it has been pointed out that if the present level of American industrialization were extended to the rest of the world, the accompanying increase in environmental pollution would bring on another Ice Age, as the massive rise in smoke and dust in the atmosphere would diminish sunlight and produce a significant lowering of the earth's temperature. An even greater danger would be the depletion of the world's oxygen supply, caused by increased chemical poisoning of the oceans.[15]

The world's current state of ecological deterioration is such as to evoke in us largely unconscious anxieties of different varieties that are consistent with those characteristic of various stages in an individual's ego-developmental history. Thus, the general apathy that I postulate is based upon largely unconscious ego defences against these anxieties. I shall speak of those ego defences associated with (a) phallic and Oedipal levels of development, (b) the earlier era coinciding with, in Kleinian terms, the depressive position, and (c) the still earlier era coinciding with the paranoid position.[16]

Phallic and Oedipal Levels

First, it is apparent in how moralistic a spirit most communications on this subject are conveyed: the speaker or writer tells us from a morally superior – and therefore safe – position, projecting his own Oedipal guilt onto us, declaring that we have raped Mother Earth and are now being duly strangled or poisoned, as if by a vengeful Jehovah, for our sin. Second, we are made to feel that the ecologists are calling upon us to relinquish our hard-won genital primacy – symbolized by our proudly cherished but ecologically offensive automobile – and return to a state of childhood, when genital mastery was something longed for but not yet achieved. Our apathy includes an unconscious, defiant refusal to do this.

Third, our fear, envy, and hatred of formidable Oedipal rivals make us view with large-scale apathy their becoming polluted into extinction. This defensive state is supported by the relatively imperceptible nature of atmospheric pollution; relatively undetectable immediately around oneself, it becomes horrifyingly evident from a distance – as from an ascending plane – as something attacking and enshrouding them, all the others with the exception of oneself, including, of course, one's Oedipal rivals. Freud, it seems to me, gave us to understand that the Oedipal struggle, in normal development, has an innately foreclosed outcome: after much inner rage and anguish, the youth or young girl must eventually come to the realization that each parent belongs sexually to the other.[17] I think Freud greatly underestimated how formidable an Oedipal rival the son or daughter remains to both the parents, and how frequently it is the youthful contestant who, in essence, becomes the victor in the intensity of emotional attachment in the Oedipal contest. Thus, I think that one of the main reasons for fathers' relative apathy towards conditions that threaten to extinguish their sons – whether these conditions be the war in Vietnam or the growing state of environmental pollution – is that such conditions promise to extinguish an Oedipal rival never fully conquered. Our unconscious hatred of succeeding generations – of our progeny and of their progeny in turn – our vengeful determination to destroy their birthright through its neglect, in revenge for the deprivations we suffered at our parents' hands during whatever developmental era, includes and extends beyond the Oedipal conflict.

Our envy of the more favoured rival is provided vicarious satisfaction by the simple levelling effect of universal environmental pollution. The poor man can have the grim satisfaction in knowing that this pollution, to which he contributes, is menacing not only himself but also the rich man. Similarly, the majority of the earth's people who live in the undeveloped countries can see that the envied, technologically developed countries are bringing about their own downfall as part of the general ruin.

Among the many unconscious meanings that environmental pollution has for us is, I think, its externalization and reification of sexual guilt – guilt which, through these transformations, is thereby rendered more tolerable to us. The psychiatric dictionary gives only this meaning of pollution: "The discharge of semen and seminal fluid in the absence of sexual intercourse; the term is often used synonymously with nocturnal emission."[18] I surmise that the more archaic, Jehovah-like aspects of our superego so terrorize us as to render us unable to distinguish between, on the one hand, the imperceptible and inexorable ageing of our body and, on the other, the increasingly pervasive pollution of the morally pure ego ideal of our youth over the years of our adulthood and ageing. This so-called moral pollution is projected, I suggest, onto the environment, such that we feel the pure air and water, and so on, of our childhood is now lost forever. Analogously, I have the notion that the well-known pictures of the mushrooming clouds of the first atom bombs may evoke in us a near-physiological apathy that is necessary for our submission to the mushrooming, Alice in Wonderland spurts in physical growth that we cannot stop, as we are physically changed – with what may feel like explosive suddenness – from child into adult. If these surmises are valid, it is of life-and-death importance that we become aware of these differentiations. Environmental pollution is a real problem in truly outer reality, about which we are by no means powerless.

The Depressive Position

Mankind is collectively reacting to the real and urgent danger of environmental pollution much like a psychotically depressed patient bent on suicide by self-neglect – the patient who, oblivious to any urgent physical hunger, lets himself starve to death or walks uncaringly into the racing automobile traffic of a busy street. One day recently, as I was driving on the Washington Beltway, observing the general custom of travelling a few miles above the speed limit, it suddenly struck me that I was essentially hurrying to get off it – to get its murderously threatening, bleak, lonely crowdedness over with. I wondered if the same were true of most of the other drivers as well, perhaps without their realizing it. I wondered, is this not a fair reflection of how we all feel not only about the beltway but about our whole current life as it stands? Is the general apathy in the face of pollution not a statement that there is something so unfulfilling about the quality of human life that we react, essentially, as though our lives are not worth fighting to save?

A few moments ago, I was suggesting that the fact of environmental pollution tends to shield one from becoming aware of the full depth of emotional depression within oneself; instead of feeling isolated within emotional depression, one feels at one with everyone else in a "realistically" doomed world. Pollution serves not only to foreclose the future upon progeny we unconsciously hate and envy, but also to obscure a past which we unconsciously resist remembering with poignant clarity. We equate the idealized world of our irretrievably lost childhood with a non-polluted environment. We tend erroneously to assume that nothing can be done about the pollution of the present-day environment because of our deeper-lying despair at knowing that we cannot recapture the world of our childhood, and at sensing, moreover, that we are retrospectively idealizing its deprived and otherwise painful aspects. The pollution serves to maintain an illusion in us that an unspoiled, ideal childhood is still there, still obtainable, if only we could bestir ourselves and clear away what spoils and obscures its purity. In this sense, pollutants unconsciously represent remnants of the past to which we are clinging – transference distortions which permeate our present environment, shielding us from feeling the poignancy of past losses, but, by the same token, barring us from living fully in present reality. We may not feel that we have lost the world of our childhood, but rather that, omnipotently, we have spoiled it and are choosing to go on spoiling it further through our continued pollution.

In current urban life, there is no longer the close-knit fabric of interpersonal relationships, enduring over decades, that would enable one to face and accept the losses inherent in human existence – the losses involved in the growing up and growing away of one's children, the ageing and death of one's parents, and the awareness of one's spouse's and one's own inevitable ageing and death. A technology-dominated, overpopulated world has so diminished our capacity to cope with the losses that a truly human life must entail, that we become increasingly drawn into polluting our planet enough to ensure that we shall have essentially nothing to lose in our eventual dying.

The Paranoid Position

In a monograph in 1960, I discussed the infant's subjective oneness with the non-human environment, the manifold functions this environment fulfils at various stages of normal ego development, and the distortions one finds, in these regards, in the histories and present-day ego functioning of schizophrenic patients.[19] In 1961, I described schizophrenia as serving to shield the afflicted individual from a recognition of the inevitability of death.[20]

For several years, I have spent a long day each month working as a consultant at the New York State Psychiatric Institute in New York City. One evening, a year or so ago, as I was returning by cab on the Triboro Bridge, on the way to LaGuardia to catch the shuttle plane back to Washington, I was seized by an urge to leap from the cab and hurl myself off the bridge. Such urges are no stranger to me, a sufferer since childhood from a phobia of heights. But the urge this time was particularly powerful, and the determinant I was able to glimpse this time, of this tenacious, multirooted symptom, was particularly memorable, humbling, and useful to me. I felt I had to destroy myself because I simply could not face returning to my usual life in Washington, and the reason I found it intolerable to face was that I felt so shamefully and desperately unable "simply" to face the living out of my life – the growing old and dying – the commonest, most everyday thing, so my panicky thoughts went, that nearly all people do, all, that is, with the exception of those who commit suicide or take refuge in chronic psychosis.

However unique to my own individual life history must be the pattern of determinants that give rise to my particular omnipotent urge to destroy my life rather than surrender to the eventual losing of it through living and ageing and dying, I insist that my urge is not entirely irrelevant to what transpires in my fellow human beings in general: I am convinced that each of us, in their own particular way, must cope with some such irrationally omnipotent reaction to inevitable loss.

I postulate that an ecologically healthy relatedness to our non-human environment is essential to the development and maintenance of our sense of being human, and that such relatedness has become so undermined, disrupted, and distorted – concomitant with ecological deterioration – that it is inordinately difficult for us to integrate the feeling experiences, including the losses inescapable to any full-fledged human living. Over recent decades, we have come from dwelling in an outer world in which the living works of nature either predominated or were near at hand, to dwelling in an environment dominated by a technology which is wondrously powerful and yet nonetheless dead, inanimate. I suggest that, in the process, we have come from being subjectively differentiated from, and in meaningful kinship with, the outer world to finding this technology-dominated world so alien, so complex, so awe-inspiring, and so overwhelming that we have been able to cope with it only by regressing – in our unconscious experience of it – largely to a degraded state of non-differentiation from it. I suggest, that is, that this "outer" reality is psychologically as much a part of us as its poisonous waste products are part of our physical selves.

The proliferation of technology, with its marvellously complex integration and its seemingly omnipotent dominion over nature, provides us with an increasingly alluring object upon which to project our "non-human" unconscious strivings for omnipotence; hence, we tend increasingly to identify, unconsciously, with it. Concomitantly, the more "simply human", animal-nature-based components of ourselves become increasingly impoverished (due to such factors as overpopulation, the impersonal and driven turmoil of living in a technology-dominated society, the emphasis on consuming material products, and so on), and are less and less capable of integrating our "non-human" components. More comprehensively, we become increasingly unable to consciously experience the war between the "human" and the "non-human" (autistic, omnipotence-based) aspects of ourselves as an inner emotional conflict; hence, we project this conflict onto, and thus unconsciously foster, the war in external reality between the beleaguered remnants of ecologically balanced nature and human technology, which is ravaging them.

Many aspects of the ecologically deteriorating world in which we live foster in us, at a largely unconscious level, the mode of experience seen in openly crystallized form in paranoid schizophrenia and postulated as characterizing the most threatened moments of normal infancy before the establishment of a durable sense of individuality. The pervasively and increasingly polluted world in which we live – where, as one concerned individual was hardly overstating it when he said, "Everything we breathe, eat, and drink is going to kill us" – is reacted to as being our all-permeating enemy. This tends to paralyse us into terrorized inactivity, all the more so because, in this deeply regressed mode of experience, we are not at all well differentiated from the environment; hence, we have no clearly separate self with which to wage a struggle with the "outer" threat.

At this level of primitive ego functioning, there is no differentiation between any good mother and bad mother. It should not be assumed that, even at a conscious level, we have been accustomed to regarding nature as equivalent to a good mother, now in conflict with technology as a bad mother. Nature has often been a bad mother to man, and has often been rendered hospitable to him only through the workings of our good mother, technology. Now we are told that our good mother is poisoning us, and that if we do not curb her and return nature to its unfettered state, we are lost. We have worshipped technology – and our annual gross national product, which epitomizes its growth – as a kind of god, and now we confusedly gather that we are supposed to starve this god in order to save ourselves.

A major aspect of this realistic "paranoid" threat lies in our ever-present suspense – however it fluctuates between consciousness and unconsciousness – that we might all die, within hours or even less, from undeclared nuclear warfare. As James Reston recently put it:

> The bomb and the missile gave the President a power unprecedented in the history of nations, and tipped the balance in the American Federal system away from the Congress, for the nation could be destroyed before the Congress could even meet to debate a declaration of war.[21]

The undifferentiated pervasiveness of all this menace evokes, deep within us, the frozen immobility of the child whose parent – equivalent to such godlike, vague entities as the hydrogen bomb or the awesomely powerful military-industrial complex[22] – chronically threatens violence.

The secrecy and subtlety of these threats make them thoroughly akin to those which grip a patient suffering from paranoid schizophrenia. We are told that, without our having realized it, we have been taking in all manner of poisons – many known, and many presumably still unknown. The known ones include – to mention only a few – lead (of which our body already carries one-third of a lethal dose),[23] mercury, DDT, and radioactive wastes.

An enormously important factor is that, at this level of ego dedifferentiation, we project – as does the openly schizophrenic individual – our own murderousness: our own pervasive, poorly differentiated and poorly integrated murderousness, born of our terror, deprivation and frustration, upon the hydrogen bomb, the military-industrial complex, technology, and so forth. Also, because we tend to feel that sudden death from nuclear warfare is a threat entirely beyond our control, we may prefer the slower, more controllable death that pollution offers as seemingly the only alternative. We know that pollution is a process to which we contribute daily; it is something which, however small our part, we know we actively do. On the other hand, to regard such slow strangulation as an inevitable agony is to yearn for the quick relief that nuclear warfare would bring.

At an unconscious level, we powerfully identify with what we perceive as omnipotent and immortal technology, as a defence against intolerable feelings of insignificance, deprivation, guilt, fear of death, and so on. It has been said, realistically, "When it comes to salvaging the environment, the individual is almost powerless".[24] Since the constructive goal of saving the world can be achieved only by working as one largely anonymous individual among uncounted millions, in adult concert with other citizens, it is more alluring to give oneself over to secret fantasies of omnipotent destructiveness, in identification with the forces that threaten to destroy the world. This serves to shield one from the recognition of one's own guilt-laden murderous urges, experienced as being within oneself, to destroy one's own intrapersonal and interpersonal world.

Our grandiose identification with technology is reinforced by statistics which inform us our cars collectively cause as much as 80 percent of air pollution;[25] that our waste production is increasing much faster than population growth and will double in less than eight years;[26] and that, as I mentioned earlier, we in the United States consume a vastly disproportionate share of the earth's raw materials.

In childhood, a fantasized omnipotence protected us against the full intensity of our feelings of deprivation, and now it is dangerously easy to identify with seemingly limitless technology and to fail to cope with the life-threatening scarcity of usable air, food, and water on our planet. By identifying with the rich diversity and wondrous integration of technology, we shield ourselves from feeling the full extent of the deprivation – the impoverishment – of our human lives. The Nobel bacteriologist René Dubos states, "Ecological systems can develop tolerance to pollutants but in the process they tend to lose their rich complexity and stability."[27] This is true, I believe, for man's psychological life as well.

For most of us, religion offers little hope of immortality to arm us against our fear of death, and we feel too ill assured that loved ones will be there to share our griefs, or that we will live in the memories of our survivors; we sense too little contact with the descendants who will outlive us. Our frustration at the knowledge that we are merely mortal is vastly intensified by the knowledge that we have created a technology which, seemingly omnipotent and immortal itself, has not extended our only allotted lifespan much beyond the biblical threescore years and ten. So we identify unconsciously with this technology which, being inanimate, cannot die. We find assurance that, in its versatile devouring, it has grown ever more powerful as it has leapt from feeding on coal – the stores of which are now largely depleted – to oil, the stores of which are expected to be exhausted in about another 30 years, and then to uranium.[28] We find reason to hope that, before the limited stores of uranium are gone, atomic or some still more magical power will have enabled immortal technology to leave this ravaged planet behind for limitless interplanetary homes, and we secretly nourish the hope that we shall be among the handful it brings with it. In this realm of omnipotent fantasy, in fact, Mother Earth is equivalent to all of reality, which is a drag and hindrance to our yearnings for unfettered omnipotence, and we want to be rid of it.

Omnipotence is not, however, something for which man unambivalently yearns. In a recent paper concerning my analysis of a man who showed a borderline thought disorder, I presented some examples of the data that showed not only his striving for the realization of his fantasized omnipotence, but also his fear lest this be fully realized and he be disqualified, thus, from sharing human love. It may not be at all coincidental that our world today is threatened with extinction through environmental pollution, to which we are so strikingly apathetic, just when we seem on the threshold of technologically breaking the chains that have always bound our race to this planet of our origin. While cognizant that this coincidence could be fully accounted for by the present developmental stage of technology alone, I suspect that we collectively quake lest our infantile omnipotent fantasies become fully actualized through man's becoming interplanetary and thereby ceasing to be man as we have known him, inseparable from earth. I surmise that we are powerfully drawn to suicidally polluting our planet so as to ensure our dying upon it as men, rather than existing elsewhere as – so we tend distortedly to assume – gods or robots, for example.

We project upon this ecologically deteriorating world the deepest intensities of all our potentially inner emotional conflicts – including, as I mentioned earlier, the conflict between the subjectively human and the subjectively non-human components of ourselves – and, since conflict is the essence of human life, we project in this same process, in large part, our aliveness. Thomas Wolfe was, I think, projecting upon the world his inner aliveness when he wrote in his notebooks of his struggle to find his place in the world: "What it may be finally I do not know but I must build up out of chaos a strong, sufficient inner life; otherwise I will be torn to pieces in the whirlpool of the world".[29] To react with apathy to our present, pollution-ridden "real, outer" world is, I think, equivalent to defending oneself unconsciously against the experience of becoming an individual human self – a self which, by the very nature of human living, must contain a whirlpool of emotional conflicts, at times so chaotic as to threaten the dismemberment of one's very self.

Time does not permit me to include here the necessarily detailed data from my work with depressed or schizophrenic persons that would provide at least a measure of clinical documentation for these speculations. In my records, there is relatively solid clinical evidence indicative, for example, of (a) patients identifying with deadly smog; (b) patients' typically paranoid transference to me as the personification of their unwanted-child-self, who was treated in the parental family as the essentially non-human source of all the subtle and pervasive malevolence that actually polluted the idealized family atmosphere; and (c) the link between patients' subjectively non-human components and their parents' autism – such patients hating what to us is reality, because to live in that reality they would have to relinquish the yearning to identify fully with the supposed omnipotence of their parental autism. My previously mentioned monograph on the non-human environment is filled with detailed clinical data relevant to this subject.

We live today at a time when we must save the real world, or we shall use it as the instrument for destroying us all. I think that the greatest danger lies neither in the hydrogen bomb itself nor in the more slowly lethal effect of pollution from our overall technology. The greatest danger lies in the fact that the world is in such a state as to evoke our very earliest anxieties, while at the same time offering the delusional "promise" – the actually deadly promise – of assuaging these anxieties, effacing them, by fully externalizing and reifying our most primitive conflicts that produce those anxieties. In the pull upon us to become omnipotently free of human conflict, we are in danger of bringing about our extinction.

If you have found anything at all apropos among my various remarks in this elementary first effort, then I have made my initial point – namely, that we psychoanalysts must make a real contribution, along with our colleagues in other fields of science, towards meeting the ecological crisis.

Notes

1 This chapter was first published in *Psychoanalytic Review*, 59.3 (1972), pp. 361–74.
2 Gail L. Baker, of The George Washington University, in her article "Environmental Pollution and Mental Health", which came into my hands on 28 August 1970, after my first draft of this article was written, includes a thorough review of the meagre literature on this subject by behavioural scientists, including nonanalytic psychiatrists. My review of the psychoanalytic literature of the past few years yields no article on this subject, but in April 1969 I heard a psychoanalyst, Peter A. Martin, give an informal talk, subsequently published in June 1970, entitled "The End of 'Our' World". In this he makes the following comments before turning to his main theme (one not relevant to the present paper):

> Psychiatrists are familiar with the fantasy met in the early stages of schizophrenia that the world is coming to an end. [. . .] In the second half of this century, the actual presence of this destructive potential makes "psychotic" end-of-the-world fantasies not so obviously out of touch with reality.
>
> Concern about world destruction is a common mass media subject. In such presentations we hear two popular theories of how the world will end. One is the "big bang" theory; the roar of nuclear explosions will herald the end of the world. Such forces may be considered as technological extensions into reality of the destructive impulses of the id. The second theory holds that the world will end with a whimper; predictions of overpopulation leading to famine, pollution, and an uninhabitable environment can be understood as stemming from two sources within the human being. One source might be extreme id impulses toward passivity which resist the obvious call for action to preserve the species. For those who believe in the debatable death instinct theory, such irrational inactivity could be explained in this way. The other source might be passivity or helplessness of the ego in the face of danger signals calling for preservation of the self.

The above preamble is presented to show what this chapter is not about. Martin then develops the theme of his chapter, namely the apparent drawing to a close of an era of training and practice in psychiatry in which young psychiatrists are imbued with the set of professional values to which we middle-aged psychiatrists have been devoted. "In summary", he says, "the group of psychiatrists, referred to as 'our group', is observing the end of its world".

Despite the obvious relevance to my paper of certain of his remarks which I have quoted, I have placed these in this endnote because, in the main, the nature of his paper is such as to confuse the development of my own remarks.
3 Steve Cotton, ed., *Earth Day: The Beginning. A Guide for Survival Compiled and Edited by the National Staff of Environmental Action* (Arno Press and Bantam Books, 1970), p. 112.
4 Paul R. Ehrlich, *The Population Bomb* (Ballantine Books, 1968), p. 56.
5 See Wesley Marx, *The Frail Ocean* (Ballantine Books, 1967).
6 Ehrlich, *The Population Bomb*, pp. 52–53.
7 Rachel Carson, *Silent Spring* (Fawcett World Library, 1962), p. 24.
8 See Ehrlich, *The Population Bomb*, pp. 56–57; and C. F. Wurster, Jr., "DDT Reduces Photosynthesis by Marine Phytoplankton", *Science*, 159 (1967), pp. 1474–75.
9 Cotton, *Earth Day*, pp. 118–19.
10 See Richard Curtis and Elizabeth Hogan, *Perils of the Peaceful Atom: The Myth of Safe Nuclear Power Plants* (Ballantine Books, 1970).
11 Ehrlich, *The Population Bomb*, p. 18.

12 Ehrlich, *The Population Bomb*, pp. 37 and 198.

13 See the prologue to Ehrlich, *The Population Bomb*.

14 Ehrlich, *The Population Bomb*, p. 133.

15 N. Cousins, "Needed: A New Dream", *Saturday Review*, 20 June 1970, p. 18.

16 See Melanie Klein, Paula Heimann and Roger Money-Kyrle, eds, *New Directions in Psycho-Analysis* (Basic Books, 1955).

17 Sigmund Freud, "The Ego and the Id" (1923), in *Standard Edition*, vol. 19 (Hogarth Press, 1961).

18 Leland E. Hinsie and Robert J. Campbell, *Psychiatric Dictionary*, 4th edn (Oxford University Press, 1970), p. 581.

19 Harold F. Searles, *The Non-human Environment in Normal Development and in Schizophrenia* (International Universities Press, 1960).

20 Harold F. Searles, "Schizophrenia and the Inevitability of Death", *Psychiatric Quarterly*, 35 (1961), pp. 631–65. Reprinted in *Collected Papers on Schizophrenia and Related Subjects* (Hogarth Press and the Institute of Psycho-Analysis, 1965; and International Universities Press, 1965), pp. 487–520.

21 James Reston, Article on editorial page of *The New York Times*, Sunday, 24 May 1970.

22 Sidney Lens, *The Military-Industrial Complex* (Pilgrim Press, 1970).

23 Cotton, *Earth Day*, p. 159.

24 See the preface to Cotton, *Earth Day*.

25 Cotton, *Earth Day*, p. 159.

26 Cotton, *Earth Day*, p. 205.

27 Cotton, *Earth Day*, p. 206.

28 Cotton, *Earth Day*, pp. 10–11.

29 Thomas Wolfe, *The Notebooks of Thomas Wolfe* (University of North Carolina Press, 1969). Quote is reprinted in *Newsweek*, 23 February 1970, pp. 102–3.

8 What Is Psychoanalytical Enlightenment Today?

A Culture of Care as a Response to the Individual's Violability in the Face of the Climate Crisis[1]

Christine Bauriedl-Schmidt, Markus Fellner,
Monika Krimmer, and Hans-Jürgen Wirth

Psychoanalysis sets out to enlighten us and has emancipatory potential. The focus on which aspect of the individual is to be emancipated has developed in parallel with society, and we can summarize the changes it has undergone as follows: from the repressed drive, to the repressed individual, through to the regulation of relationships between individuals. The intention was for the individual to become master of its own house, to assert itself against nature, and then finally to realize itself. The climate and environmental crisis poses a challenge to this heroic image of humans and the world at many societal levels – from the COVID-19 pandemic through to new forms of right-wing populism – and psychoanalysis finds itself confronted by fallible humans who are in the process of destroying the very basis of their lives. This destruction has a particularly striking impact on those who live in countries that are not wealthy, industrialized nations and therefore have the least impact on the climate crisis. The destruction is socially organized and, from a psychoanalytical perspective, is mediated by a social unconscious that is reflected in individual ideologies, discursive formations, political narratives, and general power relations.

The climate crisis affects individuals in different ways; it involves social injustice and can be read and examined as a pathology of the social.[2] Placing human and planetary vulnerability in the foreground, and drawing on the complementary relationship between Georg Hegel's theory of recognition and Emmanuel Lévinas' theory of alterity, against the backdrop of the empirical problem of the climate crisis, we propose an "ethical turn" in psychoanalysis.[3] In the opinion of the authors of this chapter, this constitutes the most advanced form of psychoanalytical theory formation in the sense of it being a socio-critical theory of the individual. The emphasis is on exposing the denial of the climate crisis and the related, implicit social injustice, on promoting, in Sally Weintrobe's words, a "culture of care", and on assuming responsibility for the violability of the individual. In this context, we need to develop a sub-heroic image of the human being – one characterized by a humble attitude towards other people and the entire

DOI: 10.4324/9781003634959-11

environment. Young people are calling for society to take responsibility of a type that goes beyond the boundaries of one's own generation and culture, and which places the individual's vulnerability, the natural basis of our lives, and our social cohesion at the forefront of psychoanalytical Enlightenment.

The Climate Crisis as a Challenge to Psychoanalytical Enlightenment

The first question is whether psychoanalytical Enlightenment has changed significantly in the first place. In Freud's day, the task and objective of psychoanalysis were initially to render the individual unconscious conscious, while the social unconscious conscious was only a secondary concern. Thus, it is notable that, to this day, the conceptualization of a "social unconscious" remains controversial.[4] Is it the sum of individual unconscious content, or is it more of an independent entity that can be construed either as essentially ahistorical (as in Jung) or as historical and discursive (for example, in Donna Orange or according to poststructuralist thought)?

Back to Sigmund Freud: for him, the focus was first and foremost on repressed sexuality and the ways in which the individual consequently became neurotic. After World War II, the psychoanalytical perspective shifted from sexuality to the mechanisms of the Ego, and to the abilities required to adapt as smoothly as possible to society's demands. Ego psychology (whose most important representative, alongside Anna Freud, was Heinz Hartmann) dominated psychoanalytical debates, especially in the United States. From the 1960s onwards, theoretical interest concentrated on identity and the self. In sum, there was a shift in emphasis from repressed drive to repressed self. In the "Age of Narcissism", the topic of finding one's true self and self-realization takes centre stage, as do the search for identity, identity diffusion, and inner coherence.[5] Today, relationship problems take the limelight and are frequently the reason for – or at least the trigger for – a desire to "do therapy". This has been expressed in psychoanalytical theory by the "relational turn" in the development of intersubjective and relational theories. Given the relational turn's roots in the psychology of the self, one could object that the ideal of autonomy is emphasized here to the detriment of closeness or intimacy. More recent research (both in terms of object relations theory and intersubjective psychoanalysis) critiques this underlying unconscious focus on narcissistic ideals, which are associated with a cultural ideal in the industrialized nations of the West and within the psychoanalytical community.

Psychoanalytical Enlightenment has hitherto proceeded from a heroic image of the human being. "Where Id was, there Ego shall be", as Freud put it. The objective was for the individual to become aware of their sexual needs, to realize themselves, and to develop a healthy sense of self-esteem. The aim was to overcome infantile dependencies and thereby to provide

Enlightenment as the emergence from self-incurred tutelage, as defined by Immanuel Kant. The psychoanalytical image of humans and the Enlightenment ideal were not naively heroic, but self-reflective and self-critical. Nevertheless, the focus was still on self-empowerment, liberation, and emancipation. The individual and their wish for self-liberation were at the centre of things. This emphasis started to change in part with the inter-subjective approaches, since the Other was now regarded as a delimiting element – an equal who had the same claims to self-realization. The individual claim to self-realization is essentially limited in intersubjective approaches, its limits set by the corresponding claims of the Other. The Other is not only an equal who has the same claims to self-realization and thus exists as a limiting factor, but also acts as the person opposite – the "you" in a dialogue that enables the constitution of the individual in the first place. Donald Winnicott's quip that the baby does not exist without the mother means, at a general level, that the individual cannot exist without the Other.

Yet despite the relational turn, drive theory (construed hermeneutically and dialectically) remains a key point of orientation for social psychology and needs to be distinguished from a purely systemic intersubjectivism that takes its cue from radical constructivism, according to which individuals are only "nodal points of social permanent communication".[6] The "pure relationship" remains the normative ideal without a concept of the individual or self-regulation without conflict.[7]

In the climate crisis, nature reappears as a recalcitrant Opposite, albeit this time one that is inexorable. Objectively, we can neither negotiate with nor converse with nature and the climate. We cannot threaten non-human nature, nor can we ask for its forgiveness. With the climate crisis, the gaze is directed in two ways: towards the inner and towards the outer world, and the difference between real fear and neurotic fear is emphasized. The question of how psychoanalysis can contribute to an appropriate response to real threats is central to the issue of psychoanalytical Enlightenment. The threat from the climate crisis does not lead directly to a "neurotic" fear, but such fear manifests indirectly through denial and/or the ideologically underpinned "defence" against the real threat and the fear focused on reality. Thus, one goal of psychoanalytical Enlightenment remains: to distinguish between neurotic fear and real fear.

When Freud postulated that psychoanalytical therapy could be compared with trying to drain the Zuiderzee, he expressed an image of the world that regards human domination over nature as desirable. psychoanalytical Enlightenment in this sense seeks to render the human "master in his own house" and (seen dialectically) thus also as master over the nature that surrounds him. This fantasy has long since been shown by psychoanalysis to be a narcissistic fantasy of omnipotence (examples being Freud's "prosthetic God" or Horst-Eberhard Richter's "God Complex") – and yet it is, all same still, implicitly innate in the psychoanalytical view of the human being. Freud's concept of overcoming nature involves a moment of domination that needs to be called into question today. At the same time, Freud was the founder of a discourse, as Michel Foucault put it, on recognizing nature in the human being. In the sense of a critical theory (of the individual), this is evident when Max Horkheimer and Theodor Adorno speak of the "remembrance of nature in the subject".[8] This is why we need a certain refocus on drive theory – even if we have distanced ourselves from its mechanistic shallowness – and from the ideology of social conformism, autarky, and mastery of the Id (nature) in the sense of radical Ego psychology.

Psychoanalytical Enlightenment worthy of its name today should, in collaboration with its sister sciences, develop a new image of the human being and the world that foregrounds human and planetary vulnerability. Vulnerability is part of humans' fundamental constitution; it has a protective function for physical and emotional survival and thus becomes a "source of productive energies and insights".[9] Psychoanalytical-therapeutic work can therefore be construed as the encounter between two vulnerable individuals.[10] This emerges even more clearly against the backdrop of the climate crisis.

Vulnerability at the Intersection of the Double Asymmetry of the Social

Vulnerability lies at the heart of two prominent traditions in psychoanalytical theory, which, on the one hand, approach it from diametrically opposed ends and, on the other, interact reciprocally: Hegel's theory of recognition and Lévinas' theory of alterity. According to contemporary social philosopher Steffen Herrmann, the two theories complement each other, and this idea can be put to fruitful use at the metapsychological level (theory of the subject) in support of the project of psychoanalytical Enlightenment/ enlightenment.

Herrmann construes there to be a "double asymmetry of the social" while conceiving of the subject as an individual in whom two equal, vulnerable actors are represented, whose thought and action are related to each other.[11] The relationship is asymmetrical, as a subject must invariably first make a communicative pitch, which entails exposing themselves to vulnerability. For Hegel, the individual's vulnerability arises from dependence on the Other's recognition. For Lévinas, the individual is affected by the Other's desire for recognition and is thus put in a situation of having to assume responsibility for the Other. While Hegel focuses on the individual's vulnerability to the Other, Lévinas considers the human condition by starting with the Other's vulnerability and the individual's resulting responsibility towards them. If people from the rich, industrialized nations of the Global North recognize the right of people from the comparatively poor countries to equality and a successful life, then this has a regulative thrust if it limits the former's perceived claim to a CO_2-intensive lifestyle, which, among other things, is anchored in an excessive, narcissistic need for recognition.

Drawing on the work of Jacques Derrida, Herrmann elaborates on Hegel and Lévinas's shared focus on the relationship between subject/individual and Other, and the work of the two can therefore be linked. The two theoretical strands need to be construed as complementary, in that they are mutually referential, "such that the one theory can first be meaningfully understood when taken together with the respective other". What we could term the double asymmetry of the social is, as Hermann suggests, essentially the "original scene" of all social relations.[12]

Recognition Theory and the "Culture of Uncare"

Hegel's well-known conception of the reciprocal relationships of recognition has been described in terms of the master/slave dialectic. Just as recognition is here a precondition for social freedom, it can also be the element that drives unfreedom. In contrast to a heroic interpretation of the master/slave relationship, in which the master is less afraid of death than the slave and thus wins the battle for recognition, Herrmann relies on a subaltern interpretation of this battle for life and death. He places the heteronomy of the slave figure at the centre of the discussion, as it embodies a decision by the subject to subjugate itself. This so-called inferior freedom is chosen because, although the relationship of recognition may be marked by contempt, it is nevertheless secure. Herrmann speaks in this context of the damaged life of the subject when, in the role of the slave, she is confronted with the contempt of the master and internalizes it in the form of self-hatred. Herrmann's interpretation makes it clear that this is not about absolute (un) freedom, but about a spectrum of freedom that the individual can use as she sees fit. The subaltern interpretation describes the profound bonding power of asymmetrical relationships of recognition, which can explain why people subjugate themselves to an ideology if this systematic subjugation offers them recognition. People reproduce the structures that are known to support negative social conditions such as the climate crisis. In order to overcome the climate crisis with a view to the quality of life of future generations, it is necessary for individuals to renounce a CO_2-intensive lifestyle driven by an ideology of growth (e.g. air travel, driving large, fast cars). Despite awareness of the climate crisis, such a sacrifice is difficult to make because individuals fear a loss of social status and recognition.

Winnicott's dialectical theory of development envisages an individual who, on the path from absolute dependency to relative independence, learns to tolerate ambivalent feelings through the transformative interaction with the two maternal object properties (the holding environment mother, and the mother as the target object of the Id drives). In terms of recognition theory, this is a development that begins with the mother's re-entry into a state of relative dependence and vulnerability, which is necessary in order to identify with the infant's needs and to be able to know very intimately how the baby feels. Winnicott suggests that there is a development towards independence by both the mother and the baby, who are inseparably linked in a relationship in which the maternal adjustment to the relationship must gradually fail in order for the infant to realize its dependence and thus be able to begin to know that the mother is an object in the outside world. The capacity for care develops from the duality with which the infant experiences the mother: as a caring object in the holding environment and as an object that survives the destructive power of the baby's Id instincts. The fear of losing the mother is modified by the infant's tender feelings towards the holding environment, both by the guilt of destructive instinctual impulses and by the experience of giving and making amends. The ability to feel concern amounts to the ability to take responsibility for one's own instincts – and for Winnicott, this is a basic element in establishing play and work, while for Lilli Gast, it enables the recognition of temporality – that what has happened is irrevocably past.[13]

Sally Weintrobe, who has her roots in object relations theory, has been a major contributor to the psychoanalytical discourse on the climate crisis for many years.[14] She places care or carelessness at the centre of a critical reflection on capitalism, neoliberalism, and its self-destructive concept of growth and/or progress: "Endless growth on a finite planet is not possible unless one thinks differently about growth".[15] She describes the self as being caught in an ambivalent conflict between carelessness (egocentrism) and caring. Both sides of the conflict are real and need to be recognized and tolerated in the inner world of the individual. For Weintrobe, the climate crisis is not primarily an ecological problem but a humanitarian crisis. She speaks in this context of a "culture of uncare", characterized by short-term thinking, instrumental reason, and a devotion to omnipotent triumph over reality. She traces the marks of a permanent, fundamental struggle between caring and uncaring imagination in politics, culture, and within individuals. Faced with the climate crisis, healthy, caring people find themselves in a "tragic position".[16] In other words, it becomes increasingly difficult to tolerate reality in light of the irreversible damage – a nightmarish insight when one considers that the earth, seen in symbolic psychodynamic terms, is also the early mother. The fulcrum for change that Weintrobe proposes is that rich people socialized in the West take responsibility – that is, relinquish their narcissistic denial (which culminates destructively and denyingly in idealization and fantasies of omnipotence) whenever reality limits their perceived claims.

The Climate Crisis Mirrored in a Pathology of the Social

Weintrobe's concept of a "culture of uncare" as the societal basis of the climate crisis points to a prominent object of psychoanalytic social psychology, namely, in Axel Honneth's words, "society's illness".[17] In his comprehensive socio-psychological study, Honneth coined the term "pathology of the social" to describe this state of affairs. In relation to the climate crisis, we can associate "social pathologies" with the political concept of climate injustice at many conceptual intersections, especially when phenomena such as exploitation, oppression, domination or racism are involved. The concept of climate injustice describes how the means for social transformation are obstructed in those parts of the world worst affected. It also describes how people in industrialized nations maintain that they are invulnerable and can thus live in a "bubble"[18] or a "psychic retreat",[19] detached from the obverse – the consequences of their own way of life.[20]

Today, psychoanalytical Enlightenment continues to be an enlightenment of denial, repression, undoing, insulation against affects, splitting, and projection, at both societal and personal levels. Denying the climate crisis in terms of its social determination, urgency, extent, and the fact that it can be politically influenced means "turning a blind eye".[21] In other words, the aforementioned aspects of the climate crisis are intellectualized, but at the same time, any conscious awareness of the corresponding affective thrust is prevented.[22] Honneth's contemporary concept of the "pathology of the social" – with its strong roots in critical theory – considers the individual person's abilities (in their respective socio-historical context) to achieve social integration, as well as their abnormalities, as expressed in the destruction of outer and inner nature with reference to intersubjective relationships of recognition.[23]

According to Herrmann, the only way to free oneself from this is through a collective and recognitive approach in which "the social orders of recognition are transformed".[24] The focus must be on enabling people to enter into relationships of responsibility, in which alternative (lived) forms of recognition can be found: "The first step toward emancipation from pathologies [. . .] consists therefore in jointly generating alternative lifeworlds, cultures, and milieus in which individuals can acquire precisely that trust in their own selves that is the precondition for the activity of critique."[25] This is also the perspective offered by Eva von Redecker, whose core recognition-theory hypothesis is that, since early Modernity, there has been a new form of property ownership that allows liberty from serfdom but through which social relations are reified in such a way that modern forms of expression of ownership-based powers of social control arise – such as racism as a sign of modern slavery. She terms this the social domination over things. Legally enshrined ownership of other people is abolished but, despite emancipation, the removal of objects does not automatically eliminate the claim to social domination over things. This claim is expressed as phantom property – a claim to domination without objects.

This phantom property is a fundamental building block of modern identities, from which a certain flexibility of identity can be derived – an individual can either have phantom property or be phantom property, that is, dispose of others or oneself.[26] The potential for social pathology can be seen in phantom property relations. These are particularly accentuated in systematic, historically rooted relations of inequality, injustice and unfreedom, all of which result in "being phantom property" coming into play, as the vulnerability of the persons concerned is particularly great in such cases.

Von Redecker's starting point for the transformation that "sets out to obstruct the destructive wrath of capitalist society" is the omnipresent changes in everyday life.[27] For von Redecker, it is not a sudden about-turn, a rupture, or a heroic sacrifice that changes the everyday patterns of our lives, but the repetition of new routines and patterns of action that are based on constant practice. Von Redecker focuses on our duty to care for the land, the foundations of life, and our fellow human beings and advocates for the development of patterns of action that are suited to forging links that go beyond the inclusion or exclusion of certain groups of people. In summary, von Redecker juxtaposes the new values with the old ones: care for property instead of domination, sharing goods instead of exploiting them, regenerative work instead of exhaustive labour, saving life instead of destroying it.

Alterity Theory and the "Ethical Turn"

The climate crisis calls on us to form theories that entail an essential change in focus from the self to the person of the Other. Gast discerns the rudiments of a psychoanalytical theory of alterity in Freud's idea of the "complex of the ancillary person", which divides into two parts: the object that can be experienced through empathy, and the thing – the Other – that is absolutely unlike the self.[28] Lévinas takes this further and proposes an especially profound and radical theory of substantive alterity, which seems to offer an appropriate socio-philosophical response to the massive political problems of the climate crisis. For Lévinas, the Other is not the individual's doppelgänger – the Otherness of the other person cannot be possessed as selfness. In terms of the formation of psychoanalytical theory, this is interesting – with the negation of the primacy of the self, the foundation of psychoanalysis is challenged. After the relational turn has deconstructed the notion of the autonomous individual, which is essentially understood as complete, and in doing so has also dissolved the individual into a relation, into the "between the individuals", Lévinas leads us back to a singular, and therefore free, intentional individual who can be construed as being responsible. The primacy of the self (and thus of the individual's claim to power) is deconstructed, but at the same time the individual remains intact because it is construed from the point of view of the Other. This makes it abundantly clear that, for Lévinas, ethics is the primary philosophical discipline, preceding even ontology and anthropology.

On the Relationship of Nature and Culture

Mainstream Western thought construes nature and culture as existing in a dualistic relationship, which has simultaneously given rise to critical attempts to overcome this dualism dialectically. We are aware that interpretations of other cultures – for example, animist or totemist peoples – are not considered in the following analysis. At this point, we would simply like to mention the work of psychoanalyst Harold Searle, who pointed to the importance of the non-human environment for human personal development, and who suggested that the phylogenetic traces of the experience of not being alive or of not being human are so existentially threatening and frightening that this experience is repressed and confined to the subconscious as soon as the conscious mind approaches these fantasies. One might also think of Devereux, who, in "From Anxiety to Method in the Behavioural Sciences", argued that positivist methods are partly shaped by a defence against fear and that scientists fear inanimate nature because it confronts them with the non-living and thus with their own death.

For the Enlightenment project of psychoanalysis and for critical theory, it has always been important to construe nature and culture dialectically in the context of a critique of reason committed to rationality. The resulting conflict between one strand of psychoanalysis, based on instinct theory, and the other, with an intersubjective orientation, is that the latter pays too little attention to nature and, by way of compensation, relies on the methodologically positivist neurosciences (over and above their productive interdisciplinary contribution) to provide foundations for the image of the human being. The result is, paradoxically, a simple reproduction of the mechanistic images that have been overcome by the intersubjective. If psychoanalysis is to remain a critical science of the individual and advance in an enlightened manner, it must continue to tread the fine line between hermeneutic methodology, interdisciplinary openness to the natural sciences, and "remembrance of nature in the subject",[29] rooted in a critique of reason – and do so with a firm eye on current empirical problems.

Thinking about nature has taken on a new dimension in the context of the climate crisis.

Alongside the Enlightenment traditions committed to rationality, we find naturalistic, holistic, and spiritual perspectives in psychological discourse on the climate crisis that are given theoretical form under the overarching concept of "deep ecology". A psychoanalytical concept of humility could be developed further and then (with due ideological caution) form the link between these discursive approaches to the climate crisis – as ideas, they underpin much of the engagement of civil society with the climate protection movement – and psychoanalytical thought. We can identify the rudiments of such a concept, for example, in Edna O'Shaughnessy's concept of gratitude or in Wilfred Bion's description of a "beam of intense darkness" that does not claim to know, remember, or understand, and through which the psychoanalyst must also endure not knowing under "fire" too and maintain the connection to the creative agency of reverie.[30] This is a sub-heroic image of the human being, in which the heroic is never completely abandoned but, in Bion's view, is in inner and external dialogue with vulnerability.

Climate Justice Construed Psychoanalytically

Against the backdrop of an "ethical turn" in philosophy, psychoanalysts have also begun to explore how this radical thinking can be used in analysis and translated into corresponding psychoanalytical concepts. Perhaps one of the most prominent of these thinkers is Donna Orange, who also addresses the climate crisis on this basis. She conceptualizes a form of relational psychoanalysis in which she formulates a strong concept of the individual, and in which the Other is not only the limit of personal claims but also the constitutive element of knowledge itself. With regard to colonialism, slavery, and racism, she seeks to develop a concept of historical unconsciousness within the discursive field of psychoanalytic social philosophy. In relation to the climate crisis, she asks what attitudes brought us into the crisis and how we conceal them from ourselves. Anthropologically and clinically, she develops her ideas by discussing the concepts of shame and envy as the conditions of consumerism and ethical blindness. She interprets the concept of a radical ethics in terms of a "social ecology" and concludes, in a manner similar to Lévinas, that "I am indeed my brother's keeper, and there is no escape." No escape from the climate crisis, no escape from the suffering and injustice that our comfortable and mindless "lifestyles" are creating.[31] Suffering and the material needs of the Other become the intelligible needs of the individual. The starting point for transformation then comes from the ability to feel shame – namely the shame experienced when we witness the victims of oppression and violence. This shame needs to be understood theoretically and must appeal to the empathy of people in the privileged industrialized nations in the face of suffering Others.

Starting with human vulnerability, we have outlined developments in psychoanalysis from Freud to intersubjectivism through the theory of recognition. The theory of alterity completes the line drawn from intersubjectivism to the ethical turn, with a strong concept of the individual, and with the climate crisis acting as an empirical catalyst for critical theory formation. Only if the climate crisis is understood as a pathology of the social, taken seriously in terms of its historicity and benchmarked against the normative thrust of general rationality, can subjective and cooperative freedom be regained.[32]

The Fridays for Future movement, a youth-led and organized global climate strike movement, has shaken us out of our stupor – not least in relation to the need to think deeply about the social meaning of generativity and to set about developing a "generative stance" that "goes hand in hand with a special form of giving and simultaneous restraint".[33] In the face of the climate crisis, psychoanalysis – as a critical science that stood for "political containment", so to speak, until the heyday of the Frankfurt School in the early 1960s – must now, more than ever, undertake the task of theorizing a radicalized form of individual responsibility, beginning with responsive care and a recognition of vulnerability.

Notes

1 The original, shorter version of this article appeared in German: C. Bauriedl-Schmidt, M. Fellner, M. Krimmer and H.-J. Wirth, "Was ist psycho-analytische Aufklärung heute? Eine Kultur der Fürsorge als Antwort auf die Verletzbarkeit des Subjekts im Angesicht der Klimakrise", *Psyche – Zeitschrift für Psychoanalyse*, 76.8 (2022), pp. 734–44.

2 See A. Honneth "Pathologies of the Social: The Past and Present of Social Phi-losophy", in *Handbook of Critical Theory*, ed. by David M. Rasmussen (Blackwell, 1996), pp. 369–98, and "Objektbeziehungstheorie und postmoderne Identität. Über das vermeintliche Veralten der Psychoanalyse", *Psyche – Z Psychoanal*, 11 (2000), pp. 1087–109.

3 See D. Orange, *Climate Crisis, Psychoanalysis, and Radical Ethics* (Routledge, 2017).

4 See Siegfried Zepf and Dietmar Seel, *Psychoanalyse und das gesellschaftlich Unbe-wusste. Eine Entmystifizierung psychoanalytischer Konzepte* (Psychosozial-Verlag, Gießen, 2020).

5 C. Lasch, *The Culture of Narcissism* (Norton, 1979).

6 M. Altmeyer, "Auf der Suche nach Resonanz. Entwurf einer Zeitdiagnose der digitalen Moderne", *Psyche – Z Psychoanal*, 73 (2019), pp. 801–25 (p. 804).

7 M. Dornes, "Die Modernisierung der Seele", *Psyche – Z Psychoanal*, 11 (2010), pp. 995–1033 (p. 997).

8 Max Horkheimer and Theodor W. Adorno, *Dialectic of Enlightenment*, trans. by E. Jephcott (Stanford University Press, 2002), p. 40.

9 R. Waldeck, "Verletzlichkeit als produktives Potenzial. Erfahrungen und Reflex-ionen im Kontext einer Krebserkrankung", *Psyche – Z Psychoanal*, 74.12 (2020), pp. 949–74 (pp. 950–51).

10 Waldeck, "Verletzlichkeit als produktives Potenzial", p. 972.

11 S. Herrmann, *Die doppelte Asymmetrie des Sozialen nach Hegel und Levinas* (Biele-feld, 2013); and "Asymmetrical Reciprocity. From Recognition to Responsibility and Back", *Metodo*, 5.1 (2017), pp. 73–97.

12 Herrmann, "Asymmetrical Reciprocity", p. 29.

13 L. Gast, "Was bedeutet: Verantwortlichkeit? Psychoanalytische Erkundungen im Vorfeld der Frage. Ein Versuch", *Psyche – Z Psychanal*, 60 (2006), pp. 57–73.

14 See L. Bayer and J. Gaines, "Engaging with Climate Change: Psychoanalytic Perspectives, London 16–17 October 2020 Tagungsbericht", *Psyche – Z Psychoa-nal* 65 (2011), pp. 464–68.

15 Sally Weintrobe, *Psychological Roots of the Climate Crisis: Neoliberal Exceptionalism and the Culture of Uncare* (Bloomsbury, 2021), p. 68.

16 Weintrobe in conversation in 2020, borrowing from Irma Brenman-Pick, "The Climate Emergency: Psychoanalytic Perspectives. Lectures at the Freud Museum London, 23.5./30.5.2020", https://www.freud.org.uk/2020/05/22/the-climate-emergency [accessed 3 November 2024].

17 A. Honneth, "The Diseases of Society. Approaching a Nearly Impossible Con-cept", *Social Research. An International Quarterly*, 81.3 (2014), pp. 683–703.

18 Orange, *Climate Crisis, Psychoanalysis, and Radical Ethics*.

19 S. Weintrobe, "The Difficult Problem of Anxiety in Thinking about Climate Change", in *Engaging with Climate Change*, ed. by Sally Weintrobe (Routledge, 2013), pp. 33–47.

20 M. Richter, "Kritik unserer zynischen Lebensweise. Warum handeln wir wider besseres Wissen? Ein Essay", *Psychoanalyse im Widerspruch: Klimawandel*, 63.1 (2020), pp. 47–64.

21 J. Steiner, "Turning a Blind Eye: The Cover Up for Oedipus", *The International Journal of Psychoanalysis*, 12 (1985), pp. 161–72, and Bayer and Gaines, "Engaging with climate change".

22 A. Honneth, "Aneignung von Freiheit. Freuds Konzeption der individuellen Selbstbeziehung" (2007), in *Pathologien der Vernunft. Geschichte und Gegenwart der Kritischen Theorie* (Suhrkamp, 2020), p. 177.

23 See A. Honneth, "A Social Pathology of Reason: On the Intellectual Legacy of Critical Theory", in *The Cambridge Companion to Critical Theory*, ed. by Fred Leland Rush (Cambridge University Press, 2004).

24 Herrmann, *Die doppelte Asymmetrie*.

25 Herrmann, p. 214.

26 E. von Redecker, *Revolution für das Leben: Philosophie der neuen Protestformen* (S. Fischer-Verlag, 2020), pp. 34–35.

27 von Redecker, p. 147.

28 Freud (1895), p. 426, quoted by Gast, "Was bedeutet", p. 64.

29 Horkheimer and Adorno, *Dialectic of Enlightenment*, p. 40.

30 E. O'Shaughnessy, "On Gratitude", in *Inquiries in Psychoanalysis. Collected Papers of Edna O'Shaughnessy*, ed. by Richard von Rusbridger (Routledge, 2015), pp. 246–58, and James Grotstein, *A Beam of Intense Darkness: Wilfred Bion's Legacy to Psychoanalysis* (Karnac, 2007).

31 Orange, *Climate Crisis, Psychoanalysis, and Radical Ethics*, p. 114.

32 See J. Habermas, "In Memoriam Alexander Mitscherlich. 1060–1063", *Psyche – Z Psychoanal*, 36.12 (1982), pp. 1059–77, and Honneth, "A Social Pathology of Reason", pp. 336–60.

33 V. King, "Generativität und die Zukunft der Nachkommen. Krisen der Weitergabe in Generationsbeziehungen", *Generativität. Gießen*, ed. by Ingrid Moeslein-Teising et al. (Psychosozial-Verlag, 2020), pp. 13–28 (p. 18).

9 Living in Climate Crisis

A Postcolonial Psychoanalytical Viewpoint

Maria Luiza Gastal

The years 2020 and 2021 were times of extreme crisis. Alongside the warnings from Climate Panel scientists, we were confronted with a virus that had a global impact. Confined to our homes, our first reactions were a mixture of fear and hope – fear of death on a massive scale, and hope that this extreme experience would lead to a radical change in attitude towards the planet, other species, and other human beings. Online videos showed scenes of people handing out food, helping elderly neighbours who could not leave their homes, and supporting those who were suffering. However, the democratic and even revolutionary potential faded as we saw poorer nations, communities, and "essential workers" unfairly exposed, while those with more resources isolated themselves in summer homes and indulged in online shopping. As vaccines were developed, we saw poor countries and continents exhaust their supplies while the United States and Europe administered booster doses. When the pandemic was over, little or none of the heralded change remained. On the contrary, the far right gained ground in several countries, with an agenda of welfare cuts, xenophobia, racism and violence. In Brazil, where I live, there was little hope for change. Despite initiatives from civil society, there was a sense that the government was celebrating death. The population was being attacked from two sides: on the one hand, by the virus and, on the other, by the authorities who were supposed to protect them from it.

Meanwhile, in a scenario that often bordered on the apocalyptic, forests were burning and cities were being flooded and destroyed across the world. We have not changed. On the contrary, the trauma of the pandemic seems to have led to phobic reactions, from an intensification of narcissism to an amplification of the death drive. Ultimately, there has been no change in attitude towards the planet and the need to find a way out of the climate crisis has become more acute.

In this chapter, I use a postcolonial viewpoint to speculate on alternative ways out of the climate crisis. Beginning with an examination of European conceptions of nature and their relation to psychoanalytic thought, I suggest that it is in what Boaventura de Sousa Santos calls the "epistemologies of the South" that we can find a counter to the monoculturalism of Western,

DOI: 10.4324/9781003634959-12

or global Northern, modernity and a way out of a crisis that focuses on the knowledge of the global South – a culmination of peoples, animals, lands, and plants that experience the repressions of the global North and the most palpable effects of climate change. Colonialism was "an epistemological domination, an extremely unequal relationship of knowledge-power",[1] which led colonized nations to have many of their forms of knowledge suppressed. The result, besides the absurd violence of this practice, was the impoverishment of human cultural life. In our current moment, when the hegemonic forms of culture are unable to emerge from the coronavirus crisis and to confront the environmental crisis that produced it in order to preserve life, perhaps it is time to examine and finally learn from the epistemologies of the global South.

Psychoanalysis and Nature: Freud and Winnicott

In *Civilization and Its Discontents* (1930), Sigmund Freud promotes a dialogue between psychoanalysis and anthropology, seeking to understand the role of culture in our relationship with the natural world. Although he broke with modern ideals of reason and individualism by asserting the primacy of the unconscious and the importance of intersubjectivity, throughout his life he consistently maintained the notion that nature is hostile to and separate from culture.[2] Despite our best efforts to master it, Freud argues, nature is a constant threat – from within, through the drives that refuse the dictates of culture, and from without, through attacks from other species and the inorganic world. Ultimately, Freudian humanity is characterized by the threat of castration. Sublimated and spiritualized, this threat of castration emerges in art, literature, and other "higher" expressions of culture, and results in an ethos that is based on a correlate of fear and human domination over nature.[3]

Freud engaged in an intense and contradictory dialogue with the biology of his time, especially in his conception of a biologically and culturally historical unconscious. He assimilated the Darwinian vision of a historically organized nature with Ernst Haeckel's theory of recapitulation, in which he had received his initial training as a natural scientist.[4] Recapitulationism is expressed not only in anatomy – the idea, proposed by Haeckel, that the stages of development through which an organism passes reflect the stages of its evolution – but also in the individual development of the unconscious, which would partially reproduce the phylogeny of the species. Stephen Jay Gould points out that Freud's recapitulationism is much more radical than Haeckel's, since he proposes a parallel between the history of the individual and that of humanity – both biological and cultural evolution would be repeated throughout the life of the human individual.[5] As Freud states in *Introductory Lectures on Psychoanalysis* (1916–1917), "[e]ach individual somehow recapitulates in an abbreviated form the entire development of the human race".[6] In terms of culture, this is teleological, as it suggests a progression of culture towards Judeo-Christianity, with signs of this movement preserved in myths, histories of civilization, and so-called "primitive" cultures. It is a narrative that also relies heavily on the biblical myth of Eden, where knowledge was lost in exchange for dominance over nature. The result is a history of civilization in which we abandon a primitive and nomadic way of life as hunter-gatherers and settle in a territory that is owned and dominated by agriculture.

Hostility and violence are much less prominent in the work of Donald Winnicott, for whom human nature is part of "natural nature" and human instincts are animal instincts. What distinguishes the human animal is the meaning that it is able to give to the experience of instincts, making an imaginative elaboration of instinctive bodily experiences.[7] For Winnicott, this is not a psychic representation of a psychosomatic source – a "concept of demarcation between the psychic and somatic", as Freud defines the drive.[8] Instead, human instincts are empirical data – bodily excitations that are given meaning and require action to be satisfied – and this imaginative elaboration of instinctual experiences is possible to a large extent due to the profound identification between mother and child.[9] In this transitional phenomena, which allows the formation of an area of illusion in which the baby believes that the breast is his creation, there is no distinction between internal and external, no "I" and no "other". To organize itself, the baby needs its instinctive (and natural) "primary creativity" and the mother/environment. In essence, the imagined object can fill the absence of the breast until the mother presents it concretely. This naive creative capacity is shaped by material reality, and several adult manifestations and expressions of culture are inherited from this area of illusion, such as the ability to be alone, play, daydream, and possess artistic creativity. Winnicott gives great importance to care, especially the care represented by the environmental provision offered by the mother and her feminine identity. On the outside, nature is an environment that cares for and meets the basic needs of the baby, and on the inside, the womb provides everything needed for life. Consequently, the environment (external nature) and instincts (internal nature) are not sources of hostility or threat. Nature is in a sense produced by culture, just as human nature produces culture. We exist as beings immersed in, and producers of, culture, which results not only from our confrontation with nature, but also from our fundamental dependence on and coexistence with it.

Those Who Were Never, in Any Sense, Modern

One of the most powerful expressions of discontent with civilization has come from the ecological movement, which attributes human suffering to our estrangement from nature and prescribes re-approximation as a cure. New ecological theories, such as those of Donna Haraway, Tim Ingold, and Bruno Latour, seek "an epistemology embodied and immanent in lifeworld contexts", in which the division between culture and nature is challenged.

This discontent also arises from the knowledge that other cultures have different visions of and relationships with nature – some are concerned with a mother/sister earth that is not "naturally" castrated, but nurturing, receptive, and welcoming, and others with a conception of natural development that reflects a belief in the biological potency of instincts. In these views of the world, the differences between internal and external, living and non-living, human and non-human, become less acute. As Bruno Latour describes in *We Have Never Been Modern* (1991), systems and networks are as "Real as Nature, narrated as Discourse, collective as Society, existential as Being".[10] He proposes a revolution: in a time of crisis for "modernity", it is time to rethink the knowledge-power, nature-*culture* divide and learn from a perspective that takes a global view, including the diverse peoples of the world.

If Freud had such an intense dialogue with the anthropology of his time, we can resume this dialogue with contemporary anthropology. Several studies and discoveries in this field have resulted in new interpretations of human history. Above all, it no longer makes sense to think of humanity's journey as being enlightenment-oriented. This is because contemporary anthropology is increasingly aware that the history of human culture is not progressive and is characterized above all by the fact that enormously diverse forms of social organization, law structures, values, ontologies, and epistemologies have always existed.

In *Inconstancy of the Indian Soul* (2011), Eduardo Viveiros de Castro reports that the Tupinambá societies that occupied the Brazilian coast at the time of the Portuguese arrival were deeply grounded in the relationship with the other, whether human or non-human, living or non-living. In these cultures, the external was in a constant process of internalization, even to the extreme of cannibalism. The Tupinambás moved in a space between the self and the other, and Viveiros de Castro argues that this transitional space is the foundation of human nature, which we maintain throughout our lives and which forms the basis of cultural experience. These "myrtle" cultures, in Viveiros de Castro's words, are different from hegemonic "marble" cultures in many ways:

> Our current idea of culture projects an anthropological landscape populated by statues of marble, not myrtle: classical museum rather than baroque garden. We understand that every society tends to persevere in its own being, and that culture is the reflexive form of this being; we think that a violent, massive pressure is necessary for it to deform and transform itself. But, above all, we believe that the being of a society is its persevering: memory and tradition are the identity marble of which culture is made. We believe, finally, that, once converted into something other than themselves, societies that have lost their tradition have no turning back. There is no going back, the previous form has been mortally wounded. [. . .] Perhaps, however, for

societies whose (in)foundation is the relationship to others, not the coincidence with themselves, none of this makes the slightest sense.[11]

In a myrtle culture, the notion of an individual identity is called into question. Myrtle, unlike marble, is an organic material that is easy to mould but that can also revert to its original shape. As a metaphor for nature and culture, it evokes a way of existing in a transitional space of adaptation, reaction, and rebellion to the world.

Other Amazonian cosmogonies, such as those of the Yawalapíti, do not recognize a separation between nature and culture. There is no division between humans and other animals, to the extent that the concept of a non-human animal does not exist. As Viveiros de Castro describes:

> The archetypes of humanity, Sun and Moon, were born from the union of the Jaguar with a human (made by the demiurge Kwamuty), and are associated with the "bichos" as opposed to fish and birds [. . .] by denying the relationship with the Jaguar and attaching themselves affectively and "specifically" to the human mother, the mythical twins proceeded against the grain of the indigenous conceptual theory, which attributes exclusively to the father the substance of the child. To deny animality by denying paternity, to attain culture by affirming maternity – this is an idea that cannot be said to be orthodoxly Freudian.[12]

If everything is impermanent, and people, animals, and things are interchangeable, we must rely on the bonds of existence, which are constituted in the world and in its origin and which sustain the world of nature and culture. Such a peculiar and distant relationship between us and nature gives rise to legal systems that are profoundly different from our own and to human-nature relations that are much less hostile.

In *The Dawn of Everything* (2021), a book that grew out of years of dialogue between British friends David Graeber, an anthropologist and activist who died in 2020, and archaeologist David Wengrow, challenges the prevailing narrative about humanity that emerged in the nineteenth century. In essence, this narrative is that Palaeolithic humans were all hunter-gatherers living in simple and egalitarian societies that only became more complex with the invention of agriculture. Drawing on recent discoveries of archaeological sites that reveal the complexity of Palaeolithic hunter-gatherer societies, the book also challenges the notion that the populations of the Americas "discovered" by Europeans were made up of savages whose social organization reflected and was a remnant of what European "humanity" had evolved from previously. The information derived from the sites of large gatherings of hunter-gatherer groups is impactful and contributes to our understanding of Amerindian societies, such as the diversity of their social organization and various economic, political, and social regimes, and how Amerindian societies were far from being "primitive and savage", as the accounts by sixteenth-century European travellers and missionaries on the west coast of Canada suggest. Such accounts mention the virtues of freedom in these societies, the reluctance of the "Indian" to let anyone fall into a state of poverty, hunger, or misery, and the ease with which outsiders taken in by "Indian" communities were accepted into prominent positions, becoming members of families or even chiefs. However, the most well-regarded values were the intensity of their social connections: mutual care, love, and, above all, happiness, which the travellers found impossible to replicate when they returned to Europe and Canada. The travellers also pointed out the enormous capacity for reasoning and debate among the "savages". Father Le Jeune, Jesuit Superior in Canada in the 1630s, observed:

> I can say in truth that, as regards intelligence, they are in no wise inferior to Europeans and to those who dwell in France. I would never have believed that, without instruction, nature could have supplied a most ready and vigorous eloquence, which I have admired in many Hurons; or more clear-sightedness in public affairs, or a more discreet management in things to which they are accustomed.[13]

Most notably up to this point, their legal systems were generally not punitive. In the example of the Wendat, an Iroquoian-speaking nation that occupied the St. Lawrence Valley and estuary to the Great Lakes region in Canada, instead of punishing the guilty, the Wendats insisted that the entire lineage or clan pay compensation. In the words of Graeber and Wengrow, "In this way, the responsibility for keeping their own under control fell on everyone."[14] According to their contemporary, Lallemant, the Wendat "captains"

> insist that the subjects provide whatever is needed; no one is obliged to do so, but those who are willing bring publicly the contribution they wish to make; it is as if they were rivalling each other over the amount of their wealth, and this is how the desire for glory and to show themselves solicitous in the face of public welfare leads them to act on such occasions.

There were chiefs, but no one was obliged to obey them – their powers of oratory and persuasion guaranteed their authority, and the community paid for individual mistakes. Even children did not have to submit.

Human history is not one long march to the European Enlightenment, and the examples given earlier illustrate why it is necessary for psychoanalysis to continue the journey begun by Freud, but also to include an analysis of social systems so different from those we have taken as a reference for the understanding of the unconscious. The consequence of this approach is a recentring of psychoanalytic thought to focus on the climate crisis produced by capitalism.

What have we lost, in this colonial journey, in terms of possible worlds, subjectivities, and ways of being? In the colonial history of Brazil and the countries that suffered under imperialism, the natives were seen as backward, primitive, and a source of slave labour. Considered by the colonizers as "rebels and lazy", the natives resisted the enslavement and dominance of the European forces and paid a high price: death, hunger, and diseases contracted from the Europeans. As a result, slavery increased in Brazil with the trafficking of people from Africa, especially Angola and Mozambique, who were seen as disposable resources to be exploited along with the colony's living beings and land. The culture of these African peoples was incorporated and mixed with that of the indigenous population and of the colonizers but was always perceived as subordinate and inferior. Our dialogue with the indigenous peoples and with the African cultures imported into Brazilian territory has always been guided by the logic of the colonizer. With "evidence" cited in scientific discourse, the "savages" were imagined as both naive and primitive, fierce and backward. This was reinforced and sustained by paternalistic discourses as well as pure brute force. The European invasion of the Americas led to the near extermination of the indigenous populations, and their myrtle cultures were crushed under the weight of the imposed marble ones.

In *Psychological Roots of the Climate Crisis* (2013), Sally Weintrobe reveals how the climate crisis stems from the ideological conviction prevalent among rich nations (and among the rich of poor nations) that the planet and everything it produces can be converted into consumer goods. Through a perverse form of meritocracy, these elites feel entitled to the planet and everything it creates and produces. However, this kind of capitalist entitlement, in which individual rights are asserted at the expense of others, is not the only way of relating to the world. Weintrobe reminds us that psychoanalysis also allows us to think about living entitlement, that includes the capacity to love, to think, and to care.[15] Contemporary anthropology and ethnography can teach us that, with the help of psychoanalysis, the remnants of myrtle cultures may help us to understand, experience, and teach more malleable ways of being.

Freud's conception of psychoanalysis took place against the background of major historical crises: the First and Second World Wars, and the Holocaust. Psychoanalysis is, in a sense, a brainchild of the permanent crisis that the human condition imposes on us, a crisis that becomes more acute at certain historical moments. This child of modernity and its values now has to deal with new crises, of which there is no shortage. Since Freud, psychoanalysts have sought to understand the dynamics of social and political catastrophes; for example, the fifth International Congress of Psychoanalysis in Budapest in 1918 discussed the traumas of war.

Psychoanalysts have focused on the consequences of postmodernity in the form of the new illnesses that come into the clinic, often bringing up the "crisis of postmodernity". Perhaps it is the case that we need to broaden our scope to welcome these different forms of subjectivation without classifying them as deviant – as human beings, we are diverse, and it is from this diversity that a new way of listening may arise, one that is open to the challenges and complexities of the contemporary world. However, it is clear that climate change is causing trauma. In a survey of 10,000 people aged 16–25 in 10 countries, 48% of the Brazilians interviewed said that climate change has a negative impact on their intention to have children – the highest proportion on the list and well above the average of 39% among the nationalities surveyed.[16]

The climate crisis offers us with an opportunity to cultivate "myrtle crops", and the task now is to take up Graeber and Wengrow's challenge by abandoning the Western colonial and paternalistic attitude that other cultures are either more backward or more innocent in favour of respectful dialogue and mutual learning. A reciprocal exchange with the epistemologies of the South allows us to listen, talk, and learn from each other. The first step, of course, is to stop the murder of the peoples who hold this knowledge. The second is to listen to them respectfully, acknowledging our differences and disagreements, but also considering what we might learn. The equal challenge is to think with and beyond Freud in order to discover what psychoanalysis has to say about the moment of crisis in which we live, to speculate on what psychoanalysts can do in the face of the concrete threat of extinction, and to examine the role of psychoanalysis in resolving the crisis. In the space of the clinic, this work happens by analysing the unconscious mechanisms that intertwine with the external world and produce new forms of psychic suffering. On a political and social level, it involves an investigation into how the unconscious functioning of the subjects of these suicidal times can offer openings for paths of change, and on a theoretical level, we must extend Freudian thought beyond his age – a time when European culture was marked by patriarchal dominance, colonialism, and racism – and maintain a cross-disciplinary connection with anthropology and ethnography. Ultimately, psychoanalysis must listen to what non-European cultures have to say.

Notes

1 B. S. Santos and M. P. Meneses, *Epistemologias do Sul* (Cortez, 2010), p. 19.
2 Sigmund Freud, "O mal-estar na civilização", in *Obras completas*, vol. 18 (P. C. Souza, Trad., Imago, 2010), pp. 13–121 (p. 29).
3 Sigmund Freud, "Novas conferências introdutórias sobre psicanálise", in *Edição standard brasileira das obras psicológicas completas de Sigmund Freud*, vol. 22 (J. Salomão, Trad., Imago, 1996), p. 131.
4 See L. Ritvo, *A influência de Darwin sobre Freud* (J. C. C. Guimarães, Trad, Imago, 1992).
5 Stephen Jay Gould, *Ontogeny and Phylogeny* (Belknap Press and Harvard University Press, 1997).

6 Freud, "Novas conferências introdutórias sobre psicanálise".

7 See Donald Winnicott, *Natureza Humana* (D. L. Bogomoletz, Trad, Imago, 1990).

8 See Freud, "Os instintos e seus destinos", in *Obras Completas*, vol. 12, pp. 38–60.

9 Leopoldo Fulgencio, "Winnicott e o abandono dos conceitos fundamentais da metapsicologia freudiana", *Livro Anual de Psicanálise*, 23 (2009), pp. 77–99 (p. 83).

10 Bruno Latour, *We Have Never Been Modern* (Harvard University Press, 1991), p. 15.

11 Eduardo Viveiros de Castro, *A inconstância da alma selvagem e outros ensaios de antropologia* (Ubu, 2017), p. 178.

12 Viveiros de Castro, *A inconstância da alma selvagem e outros ensaios de antropologia*, p. 38.

13 David Graeber and David Wengrow, *The Dawn of Everything* (Penguin, 2021), p. 45.

14 Graeber and Wengrow, *The Dawn of Everything*, p. 42.

15 See Sally Weitrobe, *Psychological Roots of the Climate Crisis* (Bloomsbury, 2021).

16 N. Hanna, *Por "ansiedade climática", os jovens estão optando por não ter filhos* (Veja, 2021), https://bit.ly/3kqtUNF [accessed 11 September 2024].

10 Stretching Horizons

Tightening Links Between Human and Non-human to Stay in the World[1]

Maria Luiza Gastal

Freud: Nature and Culture

Donald Winnicott claimed that "it cannot be too far from the truth to say that every analyst has in mind a Freud that encompasses crucial aspects of his own personal history".[2] As a psychoanalyst and biologist, I am troubled by several aspects of Sigmund Freud's work that draw on biology.

Physicality, recapitulation, instincts and their destinies, and the death drive – which today is discussed in association with the shortening of telomeres – all border on the biological sciences.[3] In the debate about the environmental crisis, two aspects seem fundamental: firstly, Freud's radical separation of nature and culture and, secondly, the idea of culture centred around the masculine and supported by the myth of a ruling father against whom we are doomed to rebel. These aspects help us to think about how we relate to the world, to life, and to what we conventionally refer to as nature. I begin my discussion with this conception of nature, before exploring Winnicott's ideas on human nature and the practices of other cultures that hold radical conceptions of what it means to be in the world. I consider the role of psychoanalysis and psychoanalysts in the serious climate crisis that we are experiencing – a crisis shaped by our way of living in the world and the need to recognize our ties to and dependence on it.

By asserting the primacy of the unconscious and the importance of inter-subjectivity, Freud breaks with modern ideals of reason and individualism. However, he retains the modern notion of nature as hostile to and separate from culture.[4] In *Civilization and its Discontents* (1930), Freud presents nature as one of the main causes of humanity's suffering, responsible for two of the three sources of hardship: "the prepotency of nature, the fragility of our bodies, and the insufficiency of the norms that regulate human bonds in the family, the state, and society".[5] We can only acknowledge the first two causes of suffering with resignation as, despite our efforts to dominate nature, it always threatens us. The fragility of our bodies is confirmed by the inevitability of death, a limit that demonstrates how culture serves as a means of alleviating the suffering caused by the cruelty of nature. Yet the fact that culture is our weapon against nature makes it all the more painful

DOI: 10.4324/9781003634959-13

to realize that it is also a source of this suffering. The transformation of nature into culture, or the control of nature by culture, takes place through the repression and sublimation of instinctual forces. Until the end of his life, Freud maintained that there is a conflict between nature and culture – an idea that is fundamental to modern thought.[6] Aggressive impulses (part of phylogenetic nature) underpin the third cause of suffering – intersubjectivity. As natural and destructive human tendencies, aggressive impulses represent the greatest obstacle to civilization. Consequently, psychoanalysis is grounded in the idea that nature causes suffering and pain. If civilization does not achieve its goal of providing us with the happiness we desire, this is largely due to external nature, with its obstacles and deficiencies, and our internal nature, from which the destructive power of aggressiveness arises.

The phylogenetic roots of the Freudian unconscious are related to the narrated history of the human species, both as an animal and as a cultural being. Freud was trained in biology by Ernst Brücke and Carl Claus, in the tradition founded by Ernst Haeckel and under the strong influence of Jean-Baptiste Lamarck.[7] The Freudian unconscious partially reproduces, in its ontogeny, the evolution of the species. Although this recapitulationism differs from Haeckel's (such as in the possibility of different stages coexisting and earlier stages remaining repressed in the unconscious), the general theory of neurosis relies heavily on Lamarck's idea of mental recapitulation. As Stephen Jay Gould reminds us, Freud intends "nothing less than the reconstruction of human history from psychological data on the development of children and neurotics".[8] There is a parallel between individual history and human history: "the child, the modern savage, our primitive ancestor, and the adult neurotic represent the same phyletic stage – the primitive as true ancestor, the savage as modern survivor, the child as adult ancestor recapitulated in Haeckelian terms, and the neurotic as a fixed child" – by the stage of the totemic successor of the father, which is then interrupted and replaced by a stage of religious devotion.[9]

Biological as well as cultural development repeats itself throughout life. In relation to culture, this recapitulationism is also teleological – the idea that "primitive" peoples and primates of other species and genera have not yet reached our level of development. "Our" culture (European, Western, Judeo-Christian) is seen as more advanced in this journey and as a beacon for others. The signs of the inexorable – and guided – advance of culture, and its efforts to overcome the vicissitudes imposed by nature are found in myths and in the history of civilization itself – preserved "remnants" of this journey. Thus, in the totemic myth, the son kills the father, resulting in guilt and the fear of castration. In the biblical myth, Eden is lost in exchange for the knowledge required to dominate nature. In human history, therefore, people abandon a primitive way of life – nomadic, hunter-gatherer – and put down roots in a territory that begins to be deforested, controlled, and dominated. In all these narratives, humanity leaves nature behind in the name of civilization.

The "humanity" on which Freud's theory of the unconscious dwells is underpinned by fear of castration, the primacy of the phallus, and the use of reason. The threat of castration, and the resulting instinctive renunciation, allows culture to advance. Sublimated and spiritualized, the threat of castration is also transformed into art, literature, and other "higher" expressions of culture. The very idea of castration anxiety is traced back to heredity and the inheritance of acquired characteristics – it is engraved in the memory of the human species. The biological – nature – is an important part of the unconscious, but it is "invaded" by castration anxiety which, in a circular movement, departs from the cultural to become part of the biological inheritance and returns to leave its mark on culture. The primacy of culture over nature results in an ethic based on law and human domination over nature. It is also constitutively masculine. In this reasoning, biological women are regarded as having been castrated by nature and are therefore considered incapable of possessing an advanced ethic, since their constitutive castration is seen as preventing full ethical and cultural development. Instead, Freud argues, they are endowed with a vanity that is "a belated compensation for their original sexual inferiority", and this sense of inferiority results in shame and in the belief that "women have made few contributions to the discoveries and inventions of the history of civilization".[10]

Articulating Freud's ideas in relation to aspects of the ecological movement, Carlos Carvalho reminds us that this movement was created as a vehicle for those dissatisfied with civilization, without attributing to nature the roots of human suffering. On the contrary, for the ecological movement, it is the removal of the natural that causes our suffering. The solution, therefore, is a return to nature. Thus, "the ecological subject shares at some level the belief in the possibility of healing the conflict between nature and culture that Freud identified as the source of human unhappiness in 'Unease with Civilization'".[11] New ecological epistemologies, especially in the field of anthropology, seek an ecological or environmental understanding of the world, "an epistemology embodied and immanent in the contexts of the lifeworld".[12] In this process, the very division between nature and culture, and between the natural and the human, is called into question.

Ninety years after the publication of *Civilization and its Discontents*, and in the face of an environmental crisis that threatens the future of all humanity, we can pose some questions to psychoanalysis and to psychoanalysts: are there other ways of psychoanalytically conceiving the relationship between nature and culture? What can other cultures tell us about the human unconscious and our ways of inhabiting the world?

Winnicott: Human Nature Is Natural

Winnicott can help us to think about our relationship with nature from an ethical perspective informed by a feminine contribution – an ethics of care. Before undertaking medical training in London, he studied biology at Jesus College, Cambridge, and his ideas have strong biological roots.[13] His conception of natural development reflects a belief in the potency of instincts and biology. Human existence itself is structurally psychosomatic, he argues, and the substrate of the body that allows the psyche to develop is inseparable from it. The basis of the psyche is the body and, in evolutionary terms, the body was the first to develop. The psyche begins as an imaginative elaboration of the somatic functions, and its most important task is the interconnection of past experiences with potentialities, awareness of the present moment, and expectations for the future.[14]

Winnicott's conception of nature has important distinctions from Freud's. One aspect, highlighted by Leopoldo Fulgencio, is the radical difference between the Freudian concept of drive and the Winnicottian concept of instinct. While, for Freud, the drive is a psychic representation of a psychosomatic source – a "limit concept between the somatic and the psychic"[15] – for Winnicott, "instincts are empirical data, bodily excitations that receive a meaning and require action to be satisfied".[16] Thus, there is no difference between instinct and drive as Freud suggested, and, as Winnicott thought dynamically, it makes no sense to think in terms of pairs of instincts. Human instincts are of the same nature as those of other animals; we are not special in relation to them. What distinguishes the human animal from other animals is the fact that the human being gives a human meaning – in other words, it makes an imaginative elaboration of the instinctual experiences of the body.[17] Whereas in Freudian psychoanalysis there is a primacy of the mind over the body, Winnicott emphasizes a psychosomatic existence that is lived and signified in interpersonal relationships – either with oneself or with the environment.[18]

Transitional phenomena also reflect a different view of nature. The deep identification between mother and baby creates an area of illusion in which the baby believes that the mother's breast is its own creation. This imagined object, made possible by the baby's primary creativity, needs to be made concrete by the mother or environment at the moment it is imagined – the child needs its natural "primary creativity" and the mother or environment to organize itself accordingly. This naive creative capacity is gradually shaped by material reality, and various adult manifestations are inherited from this area of illusion: the ability to be alone, artistic creativity, play, and other cultural expressions. Moreover, for Winnicott, aggressiveness is not just a destructive force, but a creative expression of vitality. Along with eroticism, it is an impulse that animates the individual.

In *Bernardo's Book* (2010), Manoel de Barros introduces us to Bernardo, a boy from the Pantanal, whose relationship with the inner and outer world, nature and culture, living and non-living, moves from one place to another:

Bernardo is half-tree.
His silence is loud enough to bring Faraway birds
To set on his shoulder. His eye renews afternoons.

He keeps his tools in an old box:
1 dawn opener
1 rustling nail
1 river shrinker – and
1 horizon stretcher.
(Bernardo stretches horizons using three threads from cobwebs.
   Stretching guaranteed.)
Bernardo uproots nature:
His eye dilates the west.
(Can a man enrich nature with his incompleteness?)[19]

Another significant aspect of Winnicott's metapsychology is the emphasis on caregiving at two key moments. First, the environmental care offered by the mother, which can also be provided by men. This care is based on a female identification which is not based on incompleteness but on the special development of girls. Whereas for Freud the woman is a castrated man, Winnicott sees female identity as arising "from identification and rivalry with the mother and from the imaginative elaboration of the function of the specifically female genital organ".[20] As the baby's development progresses, another concern arises: the responsibility that the baby begins to assume towards its environment and mother when it is in a vulnerable position, which is the basis of our ability to care and repair as adults.

In an interview with Gabriel Bogossian, Ailton Krenak recounts the experience of Darcy Ribeiro in a Karaja community that illustrates this maternal capacity to care for, wait for, and offer holding to an immature infant:

Darcy Ribeiro once wrote about an experience he had with a Karaja woman. [. . .] He said that he was in the field, doing his work, and one day he observed a mother teaching a child to make Karaja dolls. The mother made one and the child passed it around and broke it. She made another one, the little girl walked by and broke it, and he thought, "She's going to slap that little girl." The mother did this several times, the little girl would break and she kept doing it. Then Darcy got annoyed, the Western mind couldn't take the rehearsal, and asked, "Why don't you tell her to stop it? Why do you let her break your art?" Then she said, "Because for her it's not good yet."[21]

The feeling that life is worth living comes from this dynamic – a caring environment (a good enough mother) allows the emergence of uniqueness and the possibility of happiness. Nature is not only danger and pain. The outside (external nature) is also an environment that cares for and meets the basic needs of the child. The inner (internal nature) is also an innate constitution for a full life. The environment is the place where the mother lives and, together with the child, makes possible the creative "madness" experienced between them, from which our capacity to exist as human beings immersed in and producing culture emerges. Culture is the result not only of our confrontation with nature, but also of our inescapable dependence on and coexistence with it. In a sense, nature is produced by culture, just as nature produces culture. Nature and culture are thus sister entities, intrinsically rooted in human nature, developing and sustaining themselves on the basis of an ethic of care, from which we learn that we are not necessarily weak or helpless – this is bonding.

Other Natures, Other Cultures, Other Bonds?

Viveiros de Castro discusses the role of war and cannibalism in the Tupinambá societies that occupied the Brazilian coast, noting that these practices were deeply rooted in the relationship with the other, whether human or not.

> The Tupinambá religion, rooted in the complex of warrior exocannibalism, projected a form where the socius was constituted in the relationship with the other, where the incorporation of the other depended on a coming out of oneself – the outside was in an incessant process of interiorization, and the inside was nothing but movement outward.[22]

This passage evokes Winnicott's idea of transitional space and his emphasis on conscious and unconscious human relationships in the management of instincts.[23] In Tupinambá culture, the relationship with the other is the foundation of the transitional space and the ontological basis of culture – a way of being in the world and recognizing oneself as part of it. As Peter Fonagy and Mary Target point out, "the fundamental dimension of development is not individual and subjective but shared".[24] Winnicott attributes the ability to inhabit the transitional space that is maintained throughout our lives to the "common experience among members of a group in art, religion, or philosophy".[25] Although Winnicott is interested in "cultural experience" as a transitional phenomenon, he reminds us that "in no cultural field is it possible to be original except on the basis of tradition", and that "the integration between originality and the acceptance of tradition as the basis of inventiveness" constitutes another example of the "reciprocal action between separation and union".[26] Transitional phenomena are a foundation of human nature and the basis of cultural experience – even in modern, individualistic culture. The correspondence between Winnicott's theory and the practices of the Tupinambás brings us back to the question of how the relationship with nature takes place in societies that are based precisely on their relationship with others – a radical relationship that can, or could, include cannibalism. The contemporary Western idea of culture is based on a classical museum rather than a baroque garden – every society tends to persevere in its own being, and culture is the reflexive form of this being. Thus, significant and violent pressure is necessary for culture to deform and transform itself – culture is made of marble, not myrtle. Above all, in this theory of culture, the essence of a society is seen as perseverance: memory and tradition are the marble of identity from which culture is made. This creates a sense that societies that have lost their tradition and have been transformed into something other than themselves cannot go back. However, Viveiros de Castro points out that the very notion of identity in Western civilization is challenged by societies whose foundation is based on a collective, rather than an individual, identity.[27] The wild soul, unstable and essentially linked to the other, gives rise to a subjectivity that is based on openness and is the expression of a way of being in which "it is exchange, not identity, that is the fundamental value to be affirmed", to recall James Clifford's profound reflection.[28] Relational affinity, not essential identity, is the value to be upheld.

The precedence of nature over culture (and the "superiority" of the latter over the former) does not exist in some indigenous cultures. Danowski and Viveiros de Castro present indigenous cosmogonies that believe in the existence of a humanity before the world (and therefore before what we call nature), where portions of a "primitive" go on to become biological species, natural accidents, and celestial bodies. In this example, culture produces nature, which is not a source of misfortune but a part of humanity itself. Other Amazonian cosmogonies, such as those of the Yawalapiti, clearly do not recognize a division between nature and culture. The most striking feature of the Yawalapiti taxonomy of what we would call living beings is the lack of separation between humans and other animals. There is no concept corresponding to our notion of "non-human animal". It is impossible, therefore, to make nature conform to a general idea of animality. The archetypes of humanity – Sun and Moon, for example – are believed to be born from the union of a jaguar and a human (made by the demiurge Kwamuty) and are associated with "beasts" as opposed to fish and birds. It is worth noting that by denying their relationship with the jaguar and attaching themselves affectively and "specifically" to the human mother, the mythical twins proceeded to counter the indigenous conceptual theory that attributes the substance of the child exclusively to the father. To deny animality by denying paternity, to achieve culture by affirming maternity – this is an idea that runs counter to Freudian orthodoxy.[29]

It is impossible not to think of the boy Bernardo. Everything in his environment is interchangeable and interconnected – people, animals, things. Everything in Bernardo's (and Manoel's) world is nature, including the instruments in the boy's toolbox and the walls that the speaker comes across:

Every time I find a wall
it gives me up to its slugs.

I don't know if this is a repetition of me or of the slugs. I don't know if
    this is a repetition of the walls or of me.
Am I included in the slugs or in the walls? It seems that slug is only a
    disclosure of me. I think that inside my shell
there is no animal:
It has a fierce silence.
I stretch my slug shyness until I tease the stone.[30]

We are not Tupinambás – we did not have the opportunity to be. The invasion of the Americas by the Europeans resulted in the near extermination
of native populations and the suppression of their cultures. Our original
culture made of myrtle was largely suffocated by the weight of marble, but
it is here, in the global South, that remnants of myrtle can help us understand, experience, and teach more malleable ways of existing, with the help
of poets and, hopefully, psychoanalysis.

What About Psychoanalysis?

As Winnicott taught us, when composing the "middle group" during
the years of struggle between Anna Freud and Melanie Klein at the British
Psychoanalytic Society, thinking beyond Freud is not the same as abandoning him. Rather, it is a process of extending Freudian thought beyond
his lifetime, during which patriarchy, colonialism, and racism towards
non-Europeans permeated European culture. If he were with us, Freud
would probably be thinking about the climate crisis – perhaps interested
in what myrtle cultures can teach us. However, we can now think beyond
Freud in order to consider what psychoanalysis can contribute to the crisis
we are experiencing, to speculate on what clinics can do in the face of the
concrete threat of extinction and on the role our field might play in a possible resolution of the crisis. In undertaking this task, I also hear Winnicott's
voice calling for the thought and action of psychoanalysts:

> The environment is something to which the mature individual con-
> tributes and for which they assume responsibility. A community
> with a sufficiently high proportion of mature individuals provides
> the basis for what is called democracy. If the proportion of mature
> individuals is too low, democracy cannot become a political fact since
> affairs will be swayed by the immature, that is to say, by those who
> by identification with the community lose their own individuality or
> by those who never achieve more than the attitude of the individual
> dependent upon society.[31]

Thinking about the relationship between psychoanalysis and nature, even
in ways different from those proposed by Freud, may be an ethical imposi-
tion on psychoanalysis at this moment. The data on deforestation in Brazil,

the rise in earth's temperature, the increase in extreme weather phenomena, and the hunger that devastates huge portions of the human population impose on us a commitment to reflect on our relationship with the world.

This concern can express itself in various ways. We can ask ourselves whether patients bring climate change into their analysis and whether their analysts should seek to understand the effects of the crisis on them. We can speculate whether psychoanalysis can help us understand human behaviour in the face of the climate crisis, or even whether it has the potential to contribute to solutions. From the clinical to the social, from the ethical to the political, psychoanalysis and psychoanalysts are necessarily confronted with environmental matters.

Is climate change our patients' business? European colleagues report that it is appearing more and more in analyses, as conscious or unconscious content, demanding that analysts reflect on this phenomenon in the clinic. In my clinical context, below the equator, where myrtle is embedded in marble, our patients refer to the social and natural world, bringing up issues that, although related to the climate crisis, do not explicitly appear as such. Poverty, unemployment, and violence are old scourges of Brazilian society, ever present in our consultation rooms.

Impoverished, coastal southern countries with great social inequality are the most affected by climate change, and old problems can give the false impression of being distinct from the global crisis. However, the tendency is for them to get worse and to appear more frequently in our offices. Still regarded as a privilege of "whiteness", these matters – like psychoanalysis – need to be decolonized. Whether psychoanalysts should try to understand the effects of the climate crisis on their patients is debatable, but I argue that they should pay as much attention to these issues as they have done in other serious crises of humanity. In 1918, the theme of the fifth IPA Congress in Budapest was the traumas of war. The effects of climate on individual subjectivities are already being felt in Brazil. In a survey of 10,000 people aged 16 to 25 across ten countries, 48% of Brazilians said that climate change negatively affects their intention to have children – the highest proportion on the list and above the average of 39% among the nationalities surveyed.[32]

Ultimately, psychoanalysis can help us to understand human behaviour in the face of the climate crisis and to think of solutions. Since Freud, psychoanalysts have sought to understand the dynamics of social and political crises. In 1915, Freud analysed the German Second Reich. In the late twentieth century, Hanna Segal thought about the Cold War, which she saw as sustained by schizo-paranoid anxieties, with unconscious desires for destruction and death projected onto the adversary on both sides of the dispute.[33] Later, she interpreted the Gulf War as a redirection of the schizo-paranoid hostility of the United States and its allies towards Saddam Hussein. Between Freud and Segal, and subsequently, many have drawn on psychoanalysis to think about human crises. For Michael Rustin, the dynamics of capitalism's elites, who ignore the risks of neglecting the planet, resemble what Segal describes.³⁴ They do not simply desire an increase in consumerism but seek to ensure their power over immediate rivals and others. Consumption becomes a sign of success, and the environmental crisis reflects the destructive dynamics of internal conflict described by Segal. Rustin sees the risk of exhausting the natural resources produced over millennia as an unconscious attack on future generations in an individualistic culture, where the exacerbation of narcissism leads to a weakening of the sense of commitment to human lineage and a denial of ancestral heritage.

Psychoanalysts have many resources with which to analyse the unconscious mechanisms that reinforce and reflect a social structure that seems more and more suicidal. Brazilian psychoanalysis is mature enough to engage in this task, and Brazil's position puts us before the ethical necessity of doing so. Our country is home to the largest tropical rainforest in the world but also the perpetrator of criminal attacks on the environment which are the cause of global concern. Furthermore, our myrtle cultures teach us that it is possible to conceive of a different relational dynamic of dependence on the environment, without the defensive urge to dominate and control it as if it were a dangerous object. In the clinical space, this work is undertaken by analysing the unconscious mechanisms that intertwine with the external world and produce new forms of psychic suffering. On political and social levels, investigating the unconscious functioning of subjects in these suicidal times can offer an opening for paths of change. On the theoretical plane, psychoanalysis offers a productive perspective on the relationship between human beings and the natural world. These will not be easy paths to follow, as some involve a radical change in our way of life, structural transformations in society, and an individual willingness to face enormous challenges. While work with patients can only happen in the analysis room, the passage from understanding to action must take place in the political and social realms.

Notes

1 Part of this chapter was presented at the roundtable discussion "The Ecological and Its Place in the Internal World" at the 110th Annual Congress of the American Psychoanalytical Society (APSA), in September 2021. It originally appeared as "Esticando horizontes e estreitando laços entre o humano e o não humano para permanecer no mundo", *Revista Brasileira de Psicanálise*, (2022), pp. 27–41. Translated by the author.
2 Donald Winnicott, *O gesto espontâneo* (Martins Fontes, 2017), p. xxxii.
3 Winnicott considered death drive an empty concept or of exclusively biological meaning: "In a discussion, it would not have the least utility to introduce the expression death drive, unless one goes directly back to Freud and speaks of the tendency of organic tissues to return to the inorganic state, which, as far as psychology is concerned, means absolutely nothing, except the statement of the obvious." Donald Winnicott, *Natureza Humana* (Imago, 1990), p. 160.

4 Zygmunt Baumann points out that it was the story of modernity that the book *Civilization's Wretchedness* told, even though the author preferred to speak of Kultur or civilization. See *Liquid Modernity* (Polity, 2000), p. 7.

5 Sigmund Freud, "Mal-Estar na Civilização", in *O mal-estar na civilização, novas conferências introdutórias à psicanálise e outros textos (1930–1936)* (Companhia das Letras, 2010), p. 29.

6 Z. Baumann, *O Mal-Estar da Pós-Modernidade* (Zahar, 1998).

7 See Lucille B. Ritvo, *Influência de Darwin sobre Freud* (Imago, 1992).

8 Stephen Jay Gould, *Ontogeny and Phylogeny* (Belknap Press and Harvard University Press, 1977), p. 159.

9 Sigmund Freud, "O Homem dos lobos", in *O História de uma neurose infantil ("O homem dos lobos"), além do princípio do prazer e outros textos (1917–1920)* (Companhia das Letras, 2010), p. 101.

10 Sigmund Freud, "Novas conferências introdutórias sobre psicanálise", in *Obras Psicológicas Completas de Sigmund Freud: ESB*, vol. XXII (Imago, 1996), p. 131.

11 Isabel Cristina Moura Carvalho, "Para além do mal estar da civilização: a cura ecológica e a educação da percepção", 30ª *Reunião Anped, GT22 – Educação Ambiental*, n. 22, pp. 1–34, http://www.anped.org.br/biblioteca/item/para-alem-do-mal-estar-da-civilizacao-cura-ecologica-e-educacao-da-percepcao [accessed 29 January 2020], pp. 15–16.

12 Carvalho, "Para além do mal estar da civilização: a cura ecológica e a educação da percepção", pp. 15–16.

13 See R. Rodman, "Prefácio", in *O gesto espontâneo*, ed. by Donald Winnicott (Martins Fontes, 2017), pp. xv–xxxix.

14 Donald Winnicott, *Natureza Humana* (Rio de Janeiro: Imago, 1990), p. 37.

15 Sigmund Freud, "Os instintos e seus destinos", in *Introdução ao narcisismo: estudos de metapsicologia e outros textos (1914–1916). Obras Completas*, vol. 12 (Companhia das Letras, 2010), p. 42.

16 Leopoldo Fulgencio, "Winnicott e o abandono dos conceitos fundamentais da metapsicologia freudiana", *Livro Anual de Psicanálise*, 23 (2009), pp. 77–99 (p. 83).

17 Winnicott, *Natureza Humana*, p. 58.

18 See Fulgencio, "Winnicott e o abandono dos conceitos fundamentais da metapsicologia freudiana".

   Of course, Freud's metapsychology recognizes the importance of the body. In "The Ego and The Id" he comments that the ego is bodily – it is not a surface entity but a projection of the surface, but here we are still in the topical, and not dynamic considerations that, according to Fulgencio, are treated by Freud as "myths", which assist the organization of empirical data, as in other sciences.

19 Manoel de Barros, *Poesia completa* (Leya, 2010), p. 322.

20 See Winnicott, *Natureza Humana*.

21 Ailton Krenak, *Ideas to Postpone the End of the World* (House of Anansi Press, 2020).

22 Eduardo Viveiros de Castro, *A inconstância da alma selvagem e outros ensaios de antropologia* (Ubu, 2017), p. 199.

23 See Fulgencio, "Winnicott e o abandono dos conceitos fundamentais da metapsicologia freudiana".

24 Peter Fonagy and Mary Target, "O brincar com a realidade. IV.Uma teoria sobre a realidade externa enraizada na subjetividade", *Livro Anual de Psicanálise*, XXIII (2009), pp. 131–49 (p. 133).

25 Donald Winnicott, *Brincar e a Realidade* (Imago, 1975), p. 29.

26 Winnicott, *Brincar e a Realidade*, p. 138.

27 Viveiros de Castro, *A inconstância da alma selvagem e outros ensaios de antropologia*. p. 178.

28 James Clifford, *The Predicament of Culture: Twentieth-Century Ethnography, Literature, and Art* (Cambridge University Press, 1988), p. 188.

29 Viveiros de Castro, *A inconstância da alma selvagem e outros ensaios de antropologia*, p. 38.

30 Manoel de Barros, *Poesia completa*, p. 320.

31 Winnicott, *Natureza Humana*, p. 173.

32 See Nathalie Hanna, *"Por 'ansiedade climática', os jovens estão optando por não ter filhos*. São Paulo: Revista Veja, out", https://veja.abril.com.br/saude/por-ansiedade-climatica-os-jovens-estao-optando-por-nao-ter-filhos/ [accessed 21 December 2021].

33 See Hanna Segal, "Silence Is the Real Crime", *International Review of Psychoanalysis*, 14.3 (1987), pp. 3–11.

34 See Michael Rustin, "How Is Climate Change an Issue for Psychoanalysis?", in *Engaging with Climate Change: Psychoanalytic and Interdisciplinary Perspectives*, ed. by Sally Weintrobe (Routledge, 2013).

11 Necropolitics

Lynne Zeavin

Of concern today is the global impact of the *Anthropocene*, a term used to describe the current era and the unprecedented impact of the human species on earth's biophysical systems. In his book *Slow Violence and the Environmentalism of the Poor* (2011), Rob Nixon notes that we are not only engaged in the hurtling changes of the Anthropocene, but also living through a time in which high-speed planetary modifications are accompanied by what he refers to as "rapid changes in the human cortex" – by which he at least means the ways our minds are being transformed by the internet and our increasing reliance on various technological resources and devices.[1] Our lives are lived in rhythm with fast-paced capitalism, which has led to rapid changes to the earth. Nixon implores us to pay attention to the ecological degradation around us – a degradation that is often the effect not of something cinematically dramatic, like the Twin Towers burning, but is rather off-stage, quiet, and insidiously destructive to ecosystems, to the intricate relations of plant and animal life, and to public health, over years if not decades. This is what he means by slow violence – violence that is ongoing, constant, and now becoming newsworthy – with record heating of the planet, fires, and floods that are the result of the thawing cryosphere, the effects of toxic waste, deforestation, greenhouse gases, our carbonized atmosphere, the radioactive aftermath of war, acidifying oceans, and tremendous species loss due to ravaged habitats. We are in a time of what Nixon calls "a host of other slowly unfolding environmental catastrophes" that constitute slow dyings in contrast to our obviously fast-paced world.[2] These slow dyings are such because their causalities are postponed, often for generations. Hence, disaster is not a future horizon we must urgently draw back from, but a condition we have already entered – even as corporations profit and continue to sustain it in the present.

Necropolitics is a key subject today, and the term is drawn from Achille Mbembe's book of the same name, published in 2019. One of the main theses of Mbembe's book is the treachery of colonialization and its threats to democracy along with the ways it exposes democracy as a system that exports death. Mbembe depicts the various ways that so-called advanced nations distribute the value of death. Only certain lives are grieved – non-human

DOI: 10.4324/9781003634959-14

lives, in particular, are treated as fungible and exploited for the value they produce rather than being cared for and mattering.

Mbembe writes that as humans we have

> never learned to live with all living species, have never really worried about the damage we as humans wreck on the lungs of the earth and on its body. Thus, we have never learned how to die. With the advent of the New World and, several centuries later, the appearance of the "industrialized races," we essentially chose to delegate our death to others, to make a great sacrificial repast of existence itself via a kind of ontological vicariate.[3]

He is arguing that rather than dying for ourselves – facing our own human limits and the limits of time and mortality – we export death to those deemed unworthy of life. For our purposes, this must also include the life of the planet.

Freud remarked that no one believes in their own death. In the unconscious, there is a blank space where knowledge of this one sure thing about our future should be. But, as Jacqueline Rose has written, the pandemic (which is itself a result of the climate catastrophe) has changed life forever. She writes:

> What on earth, we might then ask, does the future consist of once the awareness of death passes a certain threshold and breaks into our waking dreams? What, then, is the psychic time we are living? How can we prepare – can we prepare – for what is to come? If the uncertainty strikes at the core of inner life, it also has a political dimension. Every claim for justice relies on belief in a possible future, even when – or rather especially when – we feel the planet might be facing its demise. This is all the more visibly the case since the pandemic has allowed the wounds of racial, sexual and economic inequality in the modern world to rise mercilessly to the surface of our social arrangements for everyone, unavoidably, to see.[4]

Except – I would add – we know that not everyone looks . . .

My hope is to help, in some small way, with the process of looking, seeing, thinking, and examining – and to do so in the context of one another, because these things are nearly impossible to face alone. We need a collectivity or a group, and we need to call on one another to participate.

Notes

1 Rob Nixon, *Slow Violence and the Environmentalism of the Poor* (Harvard University Press, 2011), p. 12.
2 Nixon, p. 2.

3 Achille Mbembe, "The Universal Right to Breathe", trans. by Carolyn Shread, *Critical Inquiry*, 47 (Winter 2021), pp. 58–62 (p. 59). See also Achille Mbembe, *Necropolitics* (Duke University Press, 2019).

4 Jacqueline Rose, "Life after Death: How the Pandemic Has Transformed Our Psychic Landscape", *The Guardian*, 7 December 2021, https://www.theguardian.com/society/2021/dec/07/life-after-death-pandemic-transformed-psychic-landscape-jacqueline-rose [accessed 4 November 2024].

Part III
Mainly Nature

12 Trees and Other Psychoanalytic Matters

Lindsay L. Clarkson

Context

We live now in an era that Amitav Ghosh memorably calls *The Great Derangement* (2016). We are all aware that the unfolding ecological crisis is due to human practices and overuse of resources worldwide. Those most imminently affected by the transformation of the planet have contributed the least to this situation. In modern Western civilization, we have developed a conception of ourselves as set apart from the natural world. This split has encouraged our inattention to the toll our way of life exacts on nature and has resulted in an alienation *from a part of ourselves* that is aware of our place *within* our environment. As psychoanalysts and dynamic therapists, we are confronted with both our patients' and our own sense of dislocation and distress in response to the rapid pace of the earth's deterioration. In his 2015 Encyclical Letter, *Laudato Si'*, Pope Francis addresses the powerful influence of the dominant culture that privileges an attitude of entitlement – to control nature and extract and use resources without restraint – on the current state of the relationship between people and the natural world. The sole measure of worth in this mindset is monetary gain. He challenges us: "A certain way of understanding human life and activity has gone awry, to the serious detriment of the world around us. Should we not pause and consider this?"[1]

Although people can modify the natural world and shape virtual worlds, the elementary basis for all of our hard work is supplied by natural resources and processes. Humanity cannot exist without nature. Sanity founders if the sustaining and accompanying aspects of the diverse biological world and larger life systems are disregarded and ignored. It is difficult to face the depths of the evolving derangement, as it calls into question the central assumptions that have organized the way we in modern Western culture have lived and conceptualized the stability of our world. Such awareness can lead to anxieties or despair that is immobilizing and may induce us to turn away or tune out. As we are exposed to incontrovertible evidence of climate change, we may begin to conceive of nature as an enemy – dangerously out of control, menacing, and threatening our sovereignty. A different

DOI: 10.4324/9781003634959-16

perspective acknowledging the full range of being human includes a relationship with the natural world. Our capacity for kinship is vital to bring to light in the context of the evolving environmental catastrophe. Such awareness can strengthen us to cope with the hard facts we encounter when we face our responsibility for the alterations we have wrought upon the earth's inhabitants and systems. Our response will be more creative and flexible if we do not harden ourselves but instead reach for our own capacities for tenderness, community, and connection, extending the field of our concern to the more-than-human world.

Introduction

In this chapter, I turn our attention to the primary relationship we forge with the natural world, a relationship that contributes to our sense of being at home on earth and within ourselves. This connection is often neglected in our accounting of ourselves. In 1986, E. O. Wilson proposed the term "biophilia" that brings to life an aspect of ourselves that may be hidden in our day-to-day experience. Biophilia (the literal translation is "love of life") entails innate tendencies to seek out nature, to be responsive to living beings and life processes, and to desire an understanding of one's environment. Rather than finding our place in the community of life through a diminishment of a preferred view of human superiority, Wilson asserts that precise knowledge of other forms of life allows us to view our own lives in context. This, in turn, enriches us, making us more fully human, with greater tolerance for dissimilarity and otherness (Wilson 1986).

For most of our 200-thousand-year evolutionary history as a species, it has been essential to our survival to be acutely aware of our environmental surrounds, and our minds and sensory systems developed in this context. We are constitutionally tuned to value a wide variety of scents, sounds, flavours, sights, textures, and rhythms. Throughout the principal era of our existence as hunter-gatherers, we depended on a keen familiarity with other animals and plants, as well as their changing seasonal habits and availability, relying on such knowledge to serve as harbingers of change, whether of propitious or dangerous times to come. Such knowledge was preserved in culture, along with various symbolic representations and meanings of relationships with figures, processes, and beings in the natural world, and was passed down as vital to each person's survival and flourishing. Although many of us now live in urban environments, we retain the capacity to attend to the details of the more-than-human life around us. The ongoing natural environment continues to hold emotional meaning for us, and alert contact with it renews our own sense of liveliness. When we open ourselves to an appreciation of the living diversity of organisms and the intricate interweaving of their paths with other beings, we carry that responsiveness into our relationship with other people.

Many poets and artists have represented nature as a presence ineluctably interwoven with their internal worlds, a portrayal that closely reflects the reality of our biological lives as one species enmeshed in a complex web of interactions with other organisms. We have much to learn from poetry, art, and natural history with respect to developing an explicit awareness of the detailed ways we experience ourselves as part of nature and live *within* it. Such creative observers have retained a sensitivity to the living beings and processes around them, inhabiting an unfenced area of exploration, inspiration, awe, and repose.[2] As is true of all living relationships, the experience of engagement contains elements both good and bad. Contact with animals, trees, or the meter of a day can provide solidity, a structure and form within which to work or exercise one's creative powers. On the other hand, links with beings or processes untouched by human concerns or with life trajectories that lie beyond what we consider familiar can lead to encounters in which strangeness and differences emerge, if we look closely. We project into nature all manner of human motivation: welcoming and mild, or dangerously unknowable and threatening.[3] Two brief examples will illustrate:

In Robert Frost's poem *The Ax – helve*, a skilled French-Canadian woodsman points out to the narrator the qualities of the wood he evaluates in selecting the segment from which to form his ax handle:

He liked to have it slender as a whipstock,
Free from the least knot, equal to the strain
Of bending like a sword across the knee.
He showed me that the lines of a good helve
Were native to the grain before the knife
Expressed them, and its curves were no false curves
Put on from without. And there its strength lay
For the hard work.

The woodsman's sensory exploration of the ax-helve continues. As readers, we recognize Frost's exploration of the subject matter and sources of inspiration for his poetry – the respect for and deep knowledge of the forest, trees, wood, birds, and flora of New England that inhabit his mind and verse. There is a strength and form present in nature "equal to the strain of bending" as a poem comes into being. The poet finds containment as he works within "the lines of a good helve . . . native to the grain before the knife expressed them".[4]

A sense of menace may emerge when one opens oneself to awareness of the presence of the natural world. Paul Klee states:

In a forest, I have felt many times over that **it is not I who looked at the forest. Some days I felt that the trees were looking at me, were speaking to me. . . .** I was there, listening . . . I think that the painter must be penetrated by the universe and not want to penetrate it . . . I expect to be inwardly submerged, buried. Perhaps I paint to break out.[5]

Klee articulates a shift from the more usual perspective of the painter as a sensitive observer of the surrounding landscape; instead, it is he who is penetrated and must find his way back to separateness. What we experience when we allow ourselves to be in a state of reverie in the natural world is influenced by the state of our internal worlds, but also by the emanations from the real physical presence of the beings in our woods and fields.

In the chapter that follows, I bring into focus our capacity for kinship with the natural world as it appears in our everyday work. Although our inner object worlds determine much of our perception of the external environment, our relationship with the complex living world around us has its own presence within us. I believe that an additional richness of being can be found if we extend the field of our reverie to include attention to the distinctiveness of our reciprocal engagement with other beings and processes in our common home.

When we enter our offices and close our doors, we are aware that the natural world continues to exist outside. Unlike Thoreau's hand-constructed cabin, our consulting rooms are not porous to the elements – to the sound of a wood thrush, to the presence of butterflies, mosquitoes and mice, to the universe of Walden Pond's movement, wind, and weather altering its surface, light, and reflection. (Thoreau also heard civilization: the "fire-breathing" locomotive on the Fitchburg Line, and the church bells from neighbouring towns, if the breezes were auspicious).[6] When we secure our doors to create the privacy and spaciousness required for psychoanalytic work, it is not a given that we must lose awareness of the to and fro of our internal relationship with the natural world that carries on as we listen to our human companion. This relationship has many facets: familiarity, awe, intrigue, terror, passion, and indifference. Although we mean to work with full awareness in the here and now, we are unwittingly affected by certain constraints of our cultural surround, where a split is made between humans and nature.

In the weave of clinical material that follows, I emphasize the presence of the natural world in the analytic process in order to make the thread distinct. In ordinary analytic work, such awareness in the analyst's mind would be held as a delicate possibility, as the patient may have quite other ways of being and relationships in mind. The pace required for the apprehension of nature's unfolding processes, detailed presence, or connectivity goes against societal and economic pressures that encourage distractedness, haste, acquisition, and possession. I invite you to attend with an ear tuned to the presence of nature in our lives.

Life in the Consulting Room

My consulting room window reveals a patch of forest with a brook. Each patient finds her own way to engage with, or exclude, the background experience of the natural world. On our first encounter, G, a student in her twenties, described herself from a robot's vantage point: as an instrument to be tweaked and managed. G treated me as an automaton, a machine she sought out to adjust or eliminate faulty aspects of herself. She knew psychological terminology for feelings but had no personal experience of such things. When G mentioned "feelings", they bore no resemblance to intangible sensation or embodied experience but were instead metallic and heavy things to be moved around. I often felt purely practical myself in response. It took work to round myself to more usual human empathy.

G was uncomprehending or outright condescending when I tried to understand what she was saying to me rather than respond to her direct queries in factual terms. Occasionally, I noticed a glimmer of warmth in response to something I said that indicated that I could understand G's wariness to show me she had any life, because of the great risk this entailed.

One day, after a number of years of treatment, G came in looking blue. She happened to notice an orchid in my consulting room and commented on the warmth of its colour. Unexpectedly, she went on to survey her surroundings, newly aware of other plants. She said she expected that I thought carefully about what kind of environment would be good for my patients and how it might affect them. She implied that this was a delicate situation. Glancing outside the window, G remarked on a goldfinch perched on the balcony railing, mentioning the vibrancy of his yellow and the lightness of his movements. She told me she loved birds. Then G confided that she had a hamster herself, but she was always worried she didn't pay enough attention to it – that it was neglected. The hamster had a rotating wheel in which to run, but G didn't provide much stimulation – nothing very interesting. Worse, she was afraid she would forget to feed the hamster, and it would die. The way she spoke made it clear that she felt this was an uninterruptible cycle, and that her work with me followed a similar rotation, without growth or life.

Prior to this session, it was my impression that she had not perceived, except momentarily, the life in me, in my office or in herself. On this day, she was not only aware but also able to take in otherness: my attention to "patients", not yet specifically to her but with the possibility of a mother who might not forget to feed or pay attention. G went on to say, speaking more formally, that she had heard "that walking in green spaces might alleviate depression". She did feel better when she was outside and was very worried about being forced to live in a small, dark, confined space, as she imagined an apartment might be in the future. I thought that when she first spoke to me, she was feeling more warmth, but then she found herself talking more formally, closing down, re-entering the dark space.

G's original blindness to the natural world was a violent measure. I had the sense that, as a young child in order to survive physically in a terrible situation, she had ruptured her ability to be aware of a need for a living environment, turning to an autistic solution. A predictable, mechanical world devoid of life was more reliable than a highly charged, alternately intrusive and abandoning, live relationship. What might appear to be a callous lack of attention to the natural world is the consequence of a desperate, vicious attack on any experience of meaningful kinship. The starvation of the caged and robotic hamster reflected an ongoing cruelty that had created her internal poverty and despair. Her treatment of the hamster was both an identification with a murderous object and an identification with the helpless creature at the mercy of such a caretaker. G's dawning recognition of the life and otherness in the plants and birds was safer than extending herself to me as a trustworthy human presence, but it carried a tendency towards fuller contact and growth.

This small clinical moment illustrates that the specifics of a person's internal situation matter in understanding what appears to be disinterest in the natural world and in gaining insight into tenacious resistance to life-affirming adaptation.

The Internal and External Worlds

I have had a long-standing concern for the natural world. It had always seemed a private matter, like a love of poetry, that I know enhanced my capacity for containment and enriched my ability to listen to words, metaphor, and harmonics. More recently, I have begun to think that this affinity is not a separate concern from psychoanalysis, but for me somehow involves finding the right balance between appreciation of and attention to the inner and outer world. There are similarities: both studies involve privileging knowledge in depth. To know one mountain or meadow fully might be a lifetime's journey; there will always remain aspects that are unknown, hidden, or inconceivable. Interdisciplinary discussions with evolutionary biologists, palaeontologists, and ecologists concerned about the impact of humans on the natural world have impressed on me a perspective that humans exist as one species among many. The discussions highlighted the diminutive scope of the existence of *Homo sapiens* in the course of life on earth. Such a view is radically different from the psychoanalytic preoccupation with humans at the centre of attention (privileged and unique), disregarding our embeddedness in the natural world.

An increasing proportion of people live in a material world that appears to be created by human effort and expertise alone. Our daily urban environment has come to resemble the result of an omnipotent fantasy that we do control the natural world. Nature is ours to be used, and it can endlessly absorb whatever waste we produce. Technological miracles will result in limitless supplies of clean water, pure air, and adequate food. However, the natural world is not solely a mental construct.

"Every creature is better alive than dead, men and moose and pine-trees, and he who understands it aright will rather preserve life than destroy it," says Henry David Thoreau in *The Maine Woods* in 1864.[7] Thoreau refers to internal aliveness and vitality, not simply lifelike existence. This difference is often the most pressing matter in our clinical work. The threesome – "men, moose, and pine trees" – broadens our community and speaks up for an awareness we have lost sight of. We evolved from, belong in, and are sustained by the non-human environment. But we often restrict what we observe. Our glance remains at the surface and registers what is useful to us. Or we turn to a virtual reality that invites us to linger in a familiar setting under our dominion.

"Strange that so few ever come to the woods to see how the pine tree lives and grows and spires, lifting its evergreen arms to the light, to see its perfect success; but most are content to behold it in the shape of many broad boards brought to market, and deem *that* its true success!" writes Thoreau.[8] It is our own loss. The natural world is inherent in the constitution of our being, and we are the poorer for our disconnection. If we take stock of ourselves, we know we are not living in a benign and creative relationship with respect to our planet. This may well reflect our relationship with ourselves.

We have other capacities. Our youthful selves are reawakened if we observe children at play in the woods. We see a curiosity about the world, a wish to learn more about its inner workings, and the multiple physical and sensory ways they engage with sticks, toads, or the invitation of sheltering trees. We see the seamless incorporation of real features of the landscape into fantasy quests, and the exploration of the boundaries, the actual and the designated. We notice children's attraction to a smooth stone with a zigzag pattern or a butterfly's sheen: aesthetic delights they discover on their own. There is a particular freedom and looseness to this play, in which a child – alone or in concert with friends – can extend her knowledge of her own capabilities and inner workings. This lively attentiveness can be deformed by the pressures of societal or parental intrusion or indifference.

There are subtle ways our culture may lead to diminutions in awareness of the natural world. In a passage from *Landmarks*, Robert Macfarlane (2015) describes the excision of certain words from the Oxford Junior Dictionary in 2007 during the course of a periodic update. Terms removed included: acorn, buttercup, dandelion, fern, heather, heron, kingfisher, lark, otter, pasture. In their place, children now find definitions for: attachment, blog, bullet point, celebrity, chatroom, cut and paste. The poignancy of the loss of attention to and location of childhood within the natural world is affecting at a deep level. Although the word substitution by Oxford was not malevolent in intent, one wonders whether there was any consideration of the emotional effect on children caused by the loss of language to designate the non-human inhabitants – fellow travellers, really – in the woods and fields of the United Kingdom. Potential inquiries and research by children into their interdependence with other beings' lives, habitats, and approaches to the world are stunted, if not interrupted, by such an absence.

If we were ethologists studying an animal other than ourselves, we would be alert not only to the development of socialization and the emotional, familial, and hierarchical negotiations of being *within* a species but also to the growth in the complexity of an animal's physical, mental, and exploratory interaction with its environment. Perception of life in context is essential to vital knowledge of sources of food and water, the habits of other animals (both predator and prey), premonitory and present effects of weather, and sites of shelter or exposure. The animal's own physical traits and unique capabilities of temperament or intelligence are assessed and stretched in the course of getting to know the place where it lives. Evolutionary pressures prime an animal's aptitude and interest in the features of its environment relevant to the animal's particular way of being. Through constant interfacing with an animal's locale and its diverse inhabitants, aspects of the external world are taken in and become part of an animal's being.

I am suggesting that we, in concert with other animals, develop a branching growth in our relationship with the natural setting in which we live. Receptivity to the natural world develops in an area that is very personal and can remain in the background of awareness or be foregrounded in reverie or present engagements. The awareness of reciprocal impact can be a human-solitary, yet environs-rich engagement, or a moment of wonder or alarm shared with another person. The complex subject of our capacity to be akin seems a terrain worthy of psychoanalytic enquiry.

A Psychoanalytic Query

In an effort to understand human relatedness to the natural world, as a psychoanalyst, I have tried to use the tools I know. After Roger Money-Kyrle describes the evolution of his premises regarding the source of mental illness – and therefore the mechanism of therapeutic action – he settles on the idea of unconscious "misconceptions" and "disorientations" about truth.[9] It is through persistent allegiance to distortions of reality that we become ill. Money-Kyrle formulates the aim of psychoanalysis: "to help the patient understand, and so overcome, emotional impediments to his discovering *what he innately already knows*". Money-Kyrle selects three central facts that we find difficult to face and unconsciously delude ourselves to avoid an emotional reckoning. They are: "the recognition of the breast as a supremely good object", "the recognition of the parents' intercourse as a supremely creative act", and "the recognition of the inevitability of time and ultimately death".[10] What he omits as a basic fact of life is the recognition that the natural world supports both the mother and the parental couple, and carries deep time beyond our lifespans. The fundamental awareness of the importance of nature is present in the parents' minds, and eventually in the child's. *There is no possibility for triangular space – or any other aspect of human existence – without an encompassing, sustaining environment.*

Security, a sense of a home base, is founded on an orientation to a good internalized object, to a benign and generative parental couple. But this internalized object or couple does not exist in a vacuum; it is supported by good interactions with others and nourished by the real presence of the natural world. Andreas Malm, a historian and human ecologist, writes:

> Any toddler casting an eye towards her closest adults must grapple with the friction, the gravity, the light and the darkness, and the physicality of the objects surrounding her, and however perfectly she and her contemporaries subsequently learn to navigate, manipulate, refashion and seemingly subdue these and other aspects of nature, they cannot extricate themselves from the exterior materiality in which they once learned to walk and work. Some circumstances will never be of their making.[11]

What Malm is emphasizing here is the way the frame and structure of the real beings and processes present in the external environment are interwoven with our sensory perceptions, bodily actions, experience of the exterior world, and psychological sense of emplacement. These factors necessarily affect our relationships with other people from day one.

As human primates, we are not only alert to our internal experience, to our group's state of mind, but if we are awake (in Thoreau's sense), we also attend to the details of what is happening now in our surrounds. As psychoanalysts, we are more likely to subsume this awareness of the environment into human object relatedness, rather than inquire into the *separate* importance of the natural world for our psychic balance. It is possible that we are overwhelmed and intimidated by biodiversity and do not feel up to the task of contemplating the depth and richness of the earth, the smallness of our stature in the universe, and the contrasting scope of the damage we have done to the natural environment. We react by evading the knowledge that would lead to pain or alarm.

Bion, in *Learning from Experience* (1962), provides an understanding that helps illuminate our difficulty in accepting that we are dependent on and embedded within the natural world. He describes the terrible consequences to an infant's developing personality when the containing relationship between mother and infant is disturbed. In certain situations, either from the infant's side of innate difficulties with aggression, or from the mother's inability to engage with her infant, the baby is not helped to tolerate frustration or pain and instead goes down the path of evading reality. In the infant's unbearable agony, omnipotent thought substitutes a magically painless situation for the actual one. If one is in complete control and there is no dependence on a living other, one does not have to deal with frustration. The problem is solved – but at the expense of living out of contact with reality.

Bion describes the "effect of splitting whose object and effect is to enable the infant to obtain what later in life would be called material comforts without acknowledging the existence of a live object on which these benefits depend. . . . Envy of the breast's capacity for love, understanding, wisdom, poses a problem solved by the destruction of alpha function. . . . The need for love, understanding, and mental development is now deflected, since it cannot be satisfied, into the search for material comforts".[12] Weintrobe (2021) has written persuasively about the large-scale effects of neoliberalism in promoting our own (partial) wishes to be exceptional, omnipotent, and to live without the frustration of depending on or caring for separate other beings. When we treat nature solely as an economic commodity, we are deprived of the sensory pleasure and growth-promoting aspects of coming to know kinship and otherness in life. If we think of Thoreau, board feet are the success, not the spire of the pine in its own habitat.

We protest that we love traipsing in the woods, the sound of an owl hooting in the distance, or our walks by the sea. But is there something in the undisturbed state of the natural world – with its complex internal interactions, otherness, and inattention to us – that troubles us and rouses our envy? Rather than a sane response of gratitude for all that the environment provides, which could lead to concern and efforts at reparation, our impulse is to interfere – to set ourselves above. We spoil the life that we feel excludes us from its riches and disdain the intricacy of our biologic home. Our solutions to environmental degradation resulting from our material consumption involve omnipotent phantasies that entail further consumption and ingenious fixes created through technological finesse. Our thinking becomes simplistic. Harold Searles (1972) considered that our primitive fears of being non-human ourselves (ourselves as broad boards) interfere with our ability to embrace life in other beings.

In a contrasting mindset, engagement with the natural world can enhance our sense of being meaningfully human. Although our internal object world affects our perception of nature, there are early and deep differences between our relationship with people and our relationship with the living, more-than-human world. Nevertheless, we hold complex phantasies about the motivations of the natural world towards ourselves. Such phantasies may be an effort to make nature less other and more congruent with human inspiration. We can also recognize the *absence* of motivation – as understood along human lines – that exists in nature. Such awareness can be a relief and can bolster our appreciation of our own way of being, as opposed to a more compliant or conventional formula for the way one ought to be.

The containment provided by the non-human environment provides a space and context to live within that is relieving in its sturdiness and indifference to our individual concerns. The knowledge that the living external world persists can provide an anchoring if we inhabit an internal world unmoored by reliable objects. The resilience and interwoven complexity of the web of life in the face of our projected experiences of hate or despair can be strengthening, and inspire us with the courage to go on. The earth's non-human inhabitants provide a level of accompaniment in life – a quiet kinship paired with unlikeness – simply going on being. Our awareness of such difference can provide a respite when we are weary of our relentless consciousness. We find solace in the expansiveness of the countryside at times of loneliness or loss; this can be an actual walk in the woods or a remembered wander. Awareness of the beauty and biological complexity of nature can be sustaining when we are threatened by breakdown or the terror of our transience. As we encounter aspects of ourselves that have been lost or split off, the company of the natural world can be steadying.

Nature: An Aspect of the Psychoanalytic Frame

Recently I have been trying to observe the impact the natural world has on me as I am working. I am accustomed to listening in the here and now, but simultaneously attending to another melody has been demanding. It reminded me of the strain of trying to discern a distant and unusual bird call amidst a chorus of familiar bird sounds. The influences that I was acquainted with to facilitate balanced listening were the voices of my colleagues, my love of psychoanalytic understanding, and the pleasures of inquiring more deeply or accurately. What I began to perceive was that my awareness of the natural world formed a near-constant and balancing weight in my states of reverie – particularly relieving when I was subject to forceful and disturbing projections and pressures. The dependability of my affectionate regard for the trees around me provided an underpinning for my acceptance of my patient's disturbance, my ability to take in and understand, and my capacity to find a way to talk about terrible anxieties and perceptions. Such experiences exist in mindless forms – both extremely urgent and uniquely configured. I think such work is more tolerable for me as a person in the internal company of other humans, moose, and pine trees.

   In my situation, the window to the natural world is an essential and intrinsic aspect of the space I provide for my patients. I mean this both literally and as an awareness of a connection that I carry inside myself. Although I do not look out the window except peripherally while I am seeing patients, I know what is there. My consulting room overlooks a woodland area, with a small stream running through it. With the window open, the quiet murmur of the brook, the wind in the trees, the birds' songs, and the dappled light are part of the analytic setting. Sometimes torrential rains, thunderstorms, or heavy snow impose themselves. The changing light throughout the year marks the passage of time. I know this is a small ark of richness, as roads I cannot see form boundaries to the west and south, and houses limit the northern edge. In the fall, leaf blowers intrude and require the windows to be closed. Nonetheless, I am grateful for the life the trees contain and represent. The setting is infinitely variable – enriching and stabilizing for me as it can be for my patients. As patients become more comfortable in themselves and feel less alienated, they begin to notice the existence of life outside – often in fleeting moments, sometimes a whisper. The non-intrusive support of the natural world – uninterested in human-scale experience – allows both me and my patient the ability to move towards primitive experience, as we can find grounding (as in feet on the ground or leaning against a tree) despite the internal emotional forces of a hurricane.

The world outside the window can also be put to other uses – as a distraction from the work at hand, as a place to dissociate to, or as a site suitable for projection. The natural world can be so infiltrated by bizarre and aggressive fragments of a person's emotions and personality that it becomes useless for comfort or sane tethering. Whether insight gleaned from clinical work will be relevant to our larger societal dis-ease, I do not know. In the two brief examples and the extended vignette that follow, I will focus our inquiry on how an individual's state of mind affects their perception of the natural environment and the consequent engagement.

Mr Z: The window sash is slammed shut by Mr Z to create a claustrophobic and stifling atmosphere that accords with his internal experience. He hates any comfort he feels I derive from the ongoing world outside and feels triumphant that he has me captive. The shut window is also a communication of the hopelessness he feels about ever allowing a freer interchange with the reality of otherness; "taking in" is experienced as being taken over by.

Ms N: Ms N has often used the woods as a place to evacuate her experiencing self and emotions in small fragments, leaving her mindless in the room with me. It is such a powerful and violent projection that even her body, in the room at such times, is more a thing than a living entity. One day, she observes "a large . . .", she pauses, "bird". She says it was bizarre – the bird was flying upside down, with a wounded and misshapen wing. It appeared to be falling and its wings were flapping ineffectually. She wants to know – did I see it? She feels it was odd and did not fit in in the midst of the ordinary woods. I think this "seeing" is more of a projection of a slightly cohering sense of herself; she is in partial contact with aspects of herself that feel malformed, disoriented, and inhuman. Such awareness (however partial) is devastating. She forcefully evacuates the splintered perception by means of her eyes into the misshapen, sinking bird, a different use of her eyes than seeing. Her enquiry to me explores whether or not I can "see it" – that is, bear to take in and know the disturbing quality of her internal world. For Ms N, the small forest has no valence of comfort or belonging and is not really seen. Rather the woods are used as a site to receive the expelled aspects of her personality. The mindlessness induced by such mechanisms leaves no possibility for exploration of – or consolation from – the natural world.

Mr A: A number of years ago, Mr A, who had always been polite, said, frankly that he resented the fact that I was an analyst and a psychiatrist, as he was not. The weight of his statement was impersonal; I was divested of my particular self and became a professional title, comporting myself accordingly. For a long time, very little of what I thought or understood could be utilized by him for growth, as real exchange between separate and distinct beings was unendurable. Just as he was physically extremely sensitive to all manner of allergens or foreign substances, psychically we faced a similar dilemma.

Over time, Mr A had spoken with some real affection for his garden and its resident birds. Frequently he dismissed the pleasure he avowed that he received from his encounters with nature as an affectation of a desirable trait. At these times, his engagement with the natural world devolved into a more acquisitive situation, in which he was in possession of a wonderful, precious observation that I was to admire while being hopelessly excluded from such a rare encounter. Mr A would then speak down to me, "teaching" me about natural phenomena that he had researched. I was addressed as small and dim, made envious by his capacities. This treatment of me seemed to be a communication about his experience as a child and made me aware that for him to recognize that I had any understanding to offer was a shaming and intolerable state of affairs. Mr A's survival had depended on his ability to convincingly marvel at his fragile, easily injured parents, and to protect himself from their forceful intrusions and disparagement. He kept the violence of what he actually thought to himself.

From his perspective, what I understood was symmetrically filled with hate and designed to force a shameful awareness upon him that he was imperfect and stupid. To depend on another was to risk a disastrous encounter with an archaic figure. Better to be the one "in the know" and to forswear familiarity with softer feelings.

When one day I made a comment about the disturbance Mr A encountered when he felt he had learned something from me, he made it clear that I had misunderstood the situation. With a rueful smile, he said he was not interested in being given a good interpretation by me or being helped by me. He came to see me in order to *be* me. If he were me, he would not have to experience the pain or humiliation of being someone who was hungry or cold or needed something, nor would he have to experience envy of me. This was not total, of course; a part of Mr A *was* rueful and could be more receptive. Increasingly Mr A knew he missed a lot by severely restricting our relationship.

A few days later, as Mr A entered, he commented on a palm frond that was hanging a slight bit over the couch. He had a complex way of taking the couch, which gave a show of ease and settling. Over time, as he had become less wary, Mr A acknowledged that this artlessness was a pose. He feigned comfort to traverse the terrible danger of coming into contact with me, not me exactly. It was commonplace to encounter a "me" who in his mind might turn on him in a vicious attack just at the moment he began to feel he had a place. But if he found a "me" in a comfortable and thoughtful mindset, this too was agonizing when he was not in the same state. On this day Mr A signalled that he, good humouredly, would take the interfering leaf in stride but also more subtly that he was bothered that he was forced to accommodate the leaf in order to lie down. A smooth, mechanical transition was interrupted by a muddle that was briefly visible. I was not perturbed by my part in the infringement of the errant leaf, and actually felt a bit relieved to have some evidence of life.

Mr. A commented for the first time after many years of work on how well cared for the plants were in my consulting room, how much they had grown over his time here, and how much they were a part of his experience coming every day to talk to me. He said fondly that it was like being in a rainforest. His way of speaking was unusually personal and touching, somewhat wistful. The plants had grown and seemed to be thriving, whereas he had made do with thinner fare.

I thought it was not just the plants, well cared for, but some appreciation of a warmer, friendlier give-and-take that Mr A could now tolerate. Momentarily, he seemed to take pleasure in a diverse and foreign environment, with potential for life, other creatures, and other ways of being, whether it be in the tropics, my mind, or his internal world in my company. There was a recognition of me as an other, with a complexity of my own and an ongoing capacity for attentiveness – not just to the plants but to him. The way he was talking led me to understand that he surmised I was not too envious of the plants' growth, nor too depleted by their demands, and so could encourage them to develop. It occurred to me that he might even believe I would take pleasure in their way of being (a true observation).

Then a shift occurred. Mr A became physically stiffer. He went on to tell me about a graduate school seminar that addressed setting up one's classroom. His professor had made the point: if you are going to have plants in your classroom, then take good care of them. Their condition will give your students and their parents an impression of you and your ability to take care of your class. Mr A reported that then and there, he had decided never to have plants in his classroom, and this remained true. He said this with some firmness and evident relief that he had disposed of, not just his risk of exposure as a person who might be neglectful and incapable of nurture, but of any interest or longing for the potential richness and growth possible in the rainforest.

Briefly Mr A had felt like the beneficiary of my capacity to understand and care for him, that he himself was being tended to in a responsive, knowledgeable way, and that he felt appreciative. With a more benign experience of me, Mr A felt more tolerant of an integrated understanding of himself. He was able to value the rainforest, and include the jungly aspects of himself. Here was the potential for development, based on the internalization of a good, not ideal object. This awareness then became unbearable and he dismantled his understanding. Mr A reverted to a state of mind dominated by archaic figures who controlled all the resources. In this state the life was taken out of something he had recognized in me. Instead of a more organic connection with me or with the natural world, my cultivation of the plants now represented a false presentation, in which I was merely following a set of rules about how the "right kind of person" behaves. I was revealed to be making a show of nurture in order to influence people, to hold myself up as superior, and to display my beneficence in contrast to the depreciated others. On this day, we were now in the realm of "broad boards". Valuing the pine as a separate and worthwhile entity, going its own way as part of a balancing ecosystem, presented too painful a situation. In order to maintain his psychic equilibrium, Mr A dropped his concern for the natural world, valorizing a superior view of non-caring.

"Tree at My Window"

In this final section, we will turn our attention to the exchange between internal and external reality in psychic equilibrium. Certain poets are particularly sensitive to the alternating lift, terror, and benign accompaniment that familiarity with – and openness to – the presence of the natural world can afford. Their verse allows us to encounter, in a contained way, such a capacity in ourselves.

Robert Frost (1874–1963) was an American poet known for his sense of place. It is unexpected to find that Frost was born in California and lived there until he was ten, when his father died. It was only then that he moved with his mother and sister to the environs north of Boston. It was this New England countryside – its ways of life, farms, fields, and forests – that became his home, providing the source of his creativity and the background for his astute observations of human nature. Beginning in his twenties, Frost owned several small farms in New Hampshire and Vermont; he wrote poetry, taught in schools and universities, farmed, and raised a family. He was often depressed or physically indisposed, and had serious bouts of confusion and dislocation when he was unable to write. Despite his public persona as a country man, he was erudite, witty, and meticulously self-educated, with a serious knowledge not only of the classics and of poetry but also of botany, geology, and astronomy. He was always a walker. When he alluded to deer in the woods, birds migratory and resident, and seasonally blooming flowers, he knew what he was talking about.

He averred he could map the geographical site in the environs of his various farms in New England where the inspiration for each poem had arisen. Many poems involve the narrator in the poem in motion: wandering, or scything, or picking apples, or tracking down a fringed gentian in a distant field. The details of birdsong, scents of warmed grass, and acute observations of mosses and orchids convey to the listener a warmth and aliveness. But Frost's poetry is also infused with intimations of mortality, of painful awareness of diminishment, of loneliness, of madness, and of a lure to death. Although particularly attuned to what he called the sound of sense, Frost had a staunch affection for regular meter in his poetry. One feels the meter serves a vital boundary function, like the stone wall in *Mending Wall*.[13] The meter's structure provides a frame that forestalls a lurking chaos.

In *The Ax-Helm* (previously quoted in part), the woodsman searches for a section of wood with grain that is true. We can infer that a poet creating verse similarly searches out a subject that can be shaped along a solid, reliable structure of truth, allowing a certain safety and sanity, and bringing terrible anxieties down to size. Frost says:

> The background is hugeness and confusion shading away from where we stand into black and utter chaos; and against the background any small man-made figure of order and concentration. . . . To me any little form I assert upon it is velvet, as the saying is, and to be considered for how much more it is than nothing.[14]

This desperate image of the atmosphere surrounding Frost's creation of a poem is balanced at other moments by love.

In an extraordinary essay, *The Figure a Poem Makes* (1939), Frost reflects on the mysteries of how a poem comes to life. In the quote that follows, he emphasizes the love present in the making, in contrast to an attempt to keep inner terror at bay:

> Just as the first mystery was how a poem could have a tune in such a straightness as meter, so the second mystery is how a poem can have a wildness and at the same time a subject that shall be fulfilled. It should be of the pleasure of a poem itself to tell how it can. The figure a poem makes it begins in delight and ends in wisdom. The figure is the same for love.[15]

Robert Frost included "Tree at My Window" in *West-Running Brook* (1928).

Tree at My Window

Tree at my window, window tree,
My sash is lowered when night comes on;

But let there never be curtain drawn
Between you and me.

Vague dream-head lifted out of the ground,
And thing next most diffuse to cloud,
Not all your light tongues talking aloud
Could be profound.

But tree, I have seen you taken and tossed,
And if you have seen me when I slept,
You have seen me when I was taken and swept
And all but lost.

That day she put our heads together,
Fate had her imagination about her,
Your head so much concerned with outer,
Mine with inner, weather.[16]

In four brief quatrains, the poet conveys an intricate dance of feeling with the tree adjacent to his house. The poem illustrates the shifting states of mind the narrator encounters in his familiarity with the tree, and the part the tree plays in re-establishing the narrator's sanity. When he falters psychically, the narrator is strengthened by the tree's unwavering existence in the real world. What Frost addresses from his indoor vantage is not simply a *view* of the tree; it is a relationship with another being, one who is embraced and tested for endurance. The bond is strained and ultimately found to hold.

As the poem commences, Frost welcomes the open dialogue with the tree, implying an intimate companionship, stating, "let there never be curtain drawn between you and me". The tree outside is created in the reader's mind as one with a substance of its own – roots, a trunk, a canopy of branches and leaves. In the second stanza, the scene alters, and we find ourselves in a nightmarish dream state. The tree has lost its grounding in the outside world. It has been lifted from its solid orientation to the earth. Now, the tree has become a figment of the poet's imagination, with a "vague dream head" instead of a leafy crown, and "light tongues talking aloud", not leaves fluttering in the wind. The reader is swirled and disoriented by the altered language of being. The poet has lost his bearings, "taken and swept away and all but lost". He projects his breakdown into the tree: it in turn loses its separate, rooted existence. In a state of madness, the narrator folds in on himself, unsecured by the natural world.

In the subsequent stanza, the narrator recovers his wits. He remembers that he knows about the tree's otherness. The tree is recognized as not simply a mental construct of the poet. The poet's eye is unclouded, and he is shaken out of himself to see and witness the tree's separate existence and living place in nature. There is a relief in his observation (through the window) of the tree "taken and tossed" by *real* winds and storms. The tree's resilient survival fortifies the narrator's ability to contain his own anxieties in an identification with the tree's capacity to endure and continue to grow. The narrator believes it is possible that in some way (unspecified) the tree has also beheld his own breakdown and recovery. Frost concludes:

> That day she put our heads together,
> Fate had her imagination about her,
> Your head so much concerned with outer,
> Mine with inner, weather.

The light touch at the ending of the poem suggests that Frost is moving himself and his reader away from the seriousness of the precarious experience of himself that he has just recounted. He playfully invokes the role of Fate in the creation of the link between himself and the tree. Noting the contrasts in their ways of being, he deftly depicts his preoccupation with the state of his inner world and the tree's existence, grounded in the conditions of outer reality.

Frost alludes to the wildness and music, as well as the terror that are there to be discovered in a sustained encounter with the natural world. We are reminded when we hear his poetry of our own capacity for such relatedness. Awareness of both the state of one's internal environment and the condition of the external world leads to an appreciation of the complex, live processes linking the two, and contributes to a sense that we are firmly and lightly held. This recognition is vital to the preservation of life in the current era. As psychoanalysts and as plain citizens of the earth, can we rouse ourselves to sensitivity – to perceive more fully our place in the natural world, and to extend a hand to preserve the earth's richness?

Notes

1 Pope Francis, *Encyclical Letter Laudato Si of the Holy Father Francis on Care for Our Common Home* (Libreria Editrice Vaticana, 2015), p. 63.
2 L. L. Clarkson and S. Rockwell, "Receptivity to the Weight and Heft of the Natural World in Our Inner Selves", *Journal of the American Psychoanalytic Association*, 72 (2024), pp. 583–612.
3 K. Soper, *What Is Nature? Culture, Politics and the Non-human* (Blackwell, 1995).
4 R. Frost, "The Ax-Helve", in *Robert Frost: Collected Poems, Prose and Plays* (The Library of America, 1995).
5 M. Merleau-Ponty, *The Primacy of Perception* (Northwestern University Press, 1964), p. 167. Emphasis mine.
6 H. D. Thoreau, *Walden* (Alfred A. Knopf, 1992).
7 H. D. Thoreau, *The Maine Woods* (Penguin, 1988), p. 164.
8 Thoreau, *The Maine Woods*, p. 163.
9 R. E. Money-Kyrle, "Cognitive Development", *The International Journal of Psychoanalysis*, 49 (1968), pp. 691–98.

10 R. E. Money-Kyrle, "The Aim of Psychoanalysis", *The International Journal of Psychoanalysis*, 52 (1971), pp. 103–06 (pp. 103-04). Emphasis mine.
11 A. Malm, *The Progress of This Storm, Nature and Society in a Warming World* (Verso, 2018), pp. 158–59.
12 W. R. Bion, *Learning from Experience* (Tavistock, 1962), pp. 10–11.
13 R. Frost, "Mending Wall", in *North of Boston* (Dodd, Mead & Company, 1983).
14 R. Frost, *The Collected Prose of Robert Frost*, ed. by M. Richardson (Belknap Press of Harvard University Press, 2007), pp. 114–15.
15 R. Frost, *The Collected Prose*, pp. 131–33.
16 R. Frost, "Tree at My Window", in *West-running Brook* (Henry Holt, 1928).

13 On Healing Split Internal Landscapes[1]

Sally Weintrobe

In October 2010, the Institute of Psychoanalysis held a two-day interdisciplinary conference on "Engaging with Climate Change: Psychoanalytic Perspectives". A current view I agree with is that engaging with climate change is linked with caring for nature. In the paper I gave at the conference I argued that, as well as fearing and hating nature, we love nature and that loving nature is an ordinary part of human nature. I emphasized that engaging with nature necessitates engaging with politics, history, and culture to better understand how they shape and influence our relationships with nature. I explored ways in which capitalism in its current deregulated and global phase actively seeks to erode our loving feelings for nature and to persuade us that we are apart from nature, not part of nature.

Nature is defined in the environmental literature in many ways, and I think some of the views of nature are rather idealized. Our love of nature can be idealized too. Genuine love of nature is flawed and limited, may be non-continuous, and may be lost and found. It develops as we develop. As children, we love nature spontaneously, naturally, and fiercely – with our bodies, all our senses and orifices, and with an engaging curiosity. Nature seems not yet to have lost the quality of a "there-ness" that greets one – the quality vividly described by Arne Naess, founder of the Deep Ecology movement. We adults delight in this and may pine for our own loss of animism and freshness of engagement.

Our love of nature includes the erotic – that free play of sensual pleasure through all the senses. One strand of environmental literature openly declares our erotic love of nature. The poet, writer, and environmentalist Terry Tempest Williams describes nature as generating a sense of wonder and awe but also "peace in patterns". Relating to nature can involve states of calm and restoration resulting from sensual pleasure and has been found to be beneficial to our well-being.[2]

Our relationships with nature are complex and permeated with phantasy. We have a profound resistance to knowing the facts about nature and about our human nature – particularly the fact of physical death, and that being alive in a psychic sense depends on awareness of death and locating ourselves in time. We resist knowing about our utter dependence on and

DOI: 10.4324/9781003634959-17

indebtedness to nature. Freud noted the problem of feeling grateful toward a Mother Earth that gives us life and then takes it away. Freud linked our phantasized relationships with nature to the relationship with the mother. In this view, we can be seen, like small children, to take Mother Earth for granted – to treat her rather like an idealized breast-and-toilet mother, just there to fulfil our needs and to absorb our waste. However, I think there are limits to the extent to which nature is linked with the mother in the internal world of the psyche. Nature has its own specific and unique qualities. Also, awareness of indebtedness has a cultural aspect. There are other cultures than ours in which debt to nature is far more openly acknowledged.

In my chapter, I conceptualize nature as those psychic landscapes in which we place our internalized relationships with flora and fauna and with the biosphere itself. These relationships, based on our experiences of nature, are also subject to considerable distortion by phantasy. By land-scape I mean a place in the internal world of the psyche. I suggest that we locate the internal relationships that we arrange in space and time within imaginary landscapes. The term landscape has certain attributes and qualities that are missing in the more abstract term "place". One can feel attached to a landscape, more grounded in a landscape, and have more of a sense that that ground is shared with others. Internal representations of landscapes have their roots in the physical world, and the concept of an internal landscape reflects the way that despite our developed capacity for abstract thinking we remain – and need to remain – somewhat attached to the physical and the material. Also, in the unconscious, we have a tendency to represent and envision the internal world and its relationships in land-scapes, and landscapes figure prominently in our dream life.

Internal landscapes are not easy to think about and are largely uncon-scious. They are the psychic, imaginary places in which we relate to fam-ily, friends, social groups, society, democracy, politics, the marketplace, nature, and so on. We also relate to different aspects of ourselves within our inner psychic landscapes. Our landscapes are our settings – socially determined to a considerable degree – and we express and develop differ-ent aspects of our selves within these settings. We find ourselves already placed in landscapes, we find ourselves through landscapes, and we find ourselves in landscapes. Diversity of landscape is vital to forming an iden-tity sufficiently rich to promote well-being and the feeling of being alive as a person.

I suggest that a rich and elaborated sense of self depends not only on maintaining a diversity of inner landscapes but also on feeling at home within one's different landscapes. By "at home", I mean having a back-ground of safety that stems from a mindful authority we can identify with and depend on. A mindful presence can be parents, government, inde-pendent public radio, a nationally respected figure, or a legal code. It is also a part of the internal self. A mindful presence tries to imagine how the other feels, thereby including the other, whereas an unmindful presence may disregard or denigrate the other, leaving the other vulnerable, frightened, and excluded.[3]

It is my contention that deregulated capitalism attacks the diversity of inner landscape and, crucially, undermines our feeling of being at home in our various internal landscapes. It does so primarily by promoting and maintaining a culture of splitting and idealization, and by encouraging identification with superior, entitled in-groups, while increasing fear of being consigned to denigrated and non-entitled out-groups. It seeks what advertising calls "mind share", a situation in which our relationships in all our various landscapes become narrowed down to being dominated by feelings of superiority and entitlement to exploit and consume the other in the relationship without the psychic pain of guilt and responsibility. In doing so, it exploits the mind's natural tendency toward splitting and idealization and seduces the part of us all that seeks to feel superior and entitled to exemption from counting the cost of this.

I would like to focus on just one kind of inner landscape: the settings of our relationships with nature. An ordinary example of the effects of "mind share" here might be that of a person who loves birds and also loves walking, but who may be seduced away from loving birds into seeing themself as the owner of the most special binoculars, the best new walking poles, and possessing a superior capacity to find the best and most exotic global locations in which to walk and see birds. The net result is an increased carbon footprint that will adversely affect birds. I suggest that such a person has been both willingly and unwillingly colonized. What is particularly colonized, exploited, and deformed is their love of nature.

When we become identified with a position of entitlement to exploit the other without concern or apparent cost, the result is a split inner landscape. Concern becomes something we show only for those we assign to our in-groups – kept near us in our imagination – while those we assign to being far away from us we can more easily treat without concern. Spatially, the "far away" objects can be experienced as being in the shadows, in a forgotten place, denigrated and placed "on the other side of the tracks", or located spatially and temporally in a far-off land that may even be called the future.

What might a relatively non-split natural landscape look like? I suggest that, in such a landscape, common ground is shared between the self and other human and non-human species. Also, on this common ground feelings of empathy and solidarity with other life forms may even offer us solace in the face of death. We better appreciate that we are all, as different forms of life, equal in death. We can perhaps appreciate the enormity of our difficulty in not splitting our natural landscapes when we realize that we have no word for cross-species empathy, that it is clearly highly suspect to claim humaneness as a uniquely human trait, and that we regularly project onto animals all our most undesirable attributes. In addition, we are taught to disregard our empathic, projective imagination about animals' feelings on grounds that we are anthropomorphizing and/or have no scientific evidence. The idea that our species alone has a capacity for empathy and a moral life has been successfully challenged by the recent and growing body of evidence from evolutionary biology.[4]

As corporate capitalism has become progressively deregulated and with a global reach, attempts to encourage splitting have become more overt, aggressive, and intrusive – as have attempts to blunt our loving feelings for nature, feelings which are necessary to heal the splits. Shock-and-awe tactics are increasingly used.

One example is the way that many wildlife nature programmes (excluding David Attenborough's BBC Wildlife series) currently tend to promote identification with idealized, superior, and cut-off "hero" figures, while simultaneously blunting our feelings and disrupting our capacity to follow a coherent narrative. What is meant to be awesome in these nature programmes is not nature but the narrator. The sounds of nature are silenced by loud, often crass and anthropomorphizing music, and the programmes also intrude with sudden auditory flash noises and sudden visual shots that may include bloody body parts. Unannounced, these are shocking, and we know that in states of shock, people are more likely to identify with the aggressor. These techniques, now far more prevalent across the board in mainstream cinema and television, are particularly effective in deadening more concerned, depressive feelings and promoting a blank state of mind – one that blocks out the here and now of reality and seeks to replace it with the pumped-up, self-aggrandized "here and now" state of the dream world of regressive wish fulfilment.

Notes

1 This chapter was first published in *The Institute of Psychoanalysis*, Annual Issue 2011.
2 For the evidence, see Richard Louv, *Last Child in the Woods* (Atlantic Books, 2005).
3 When Margaret Thatcher said, "There is no such thing as Society," she was attacking a particular landscape. We were meant to feel outcasts in our identity as citizens, without a psychic home and location as social beings with civic rights and duties and responsibilities to and for each other. By contrast, when Nelson Mandela went on public radio in South Africa to praise Afrikaans culture after Afrikaans ceased to be an official language, he was mindful of how the Afrikaans people would feel outcast, and he sought to include them and restore their sense of dignity and belonging.
4 See Frans de Waal, *The Age of Empathy* (Harmony Books, 2009).

14 I Am the River . . .

Pushpa Misra

On one of my annual trips to the foothills of the Himalayas, I witnessed the gradual destruction of the forests. The denuded hills stared at me in their stark nakedness. I felt deeply saddened and utterly helpless. The feeling was no less intense than the loss of a loved one. I have developed a strong attachment to this part of the Himalayas. Each visit gives me a sense of homecoming, tranquillity, and peace.

Human beings have a strong affinity with nature, consciously or unconsciously. My interest in environmental ethics led me to study various theories on the subject. Arne Naess, the famous philosopher, environmentalist, and founder of the Deep Ecology movement, says:

> When I was nine or ten, I learned to enjoy the high mountains where my mother had a cottage. Because I had no father, the mountain somehow became my father, as a friendly, immensely powerful being, perfect and extremely tranquil. [. . .] Nature is overwhelmingly rich and good and does not impose anything upon us. We are completely free, our imagination is free.[1]

Nature is not something outside of us. It is both external and internal. The real objects are part of the external reality, but their images, coloured by our own emotions and projections, are part of our psychic reality. Natural objects become part of our personality, shaping and largely determining who we are. Objects from the natural environment in which we grew up become part of our personal identity, just like the parental images we project onto ourselves. Naess cites the example of people who have been resettled from the Arctic wilderness to so-called centres of development. "There is a consequent loss of personal identity," he says, because the social, economic, and natural settings are completely different. He continues:

> If people are relocated, or rather transported, from a steep mountainous place to the plains below, they also realize [. . .] that their home-place was a part of themselves and that they identified with features of that place.[2]

DOI: 10.4324/9781003634959-18

This is also evident in the longing, sometimes intergenerational, of displaced people for their original homes. Tibetan refugees still yearn to return to Tibet, and uprooted people from what used to be East Bengal never tire of talking about the places of their roots. I am reminded of Alex Hailey's famous novel *Roots: The Saga of an American Family* (1976), which illustrates the point very well. There are two aspects to this strong desire to be united with one's original habitat. Initially, a strong identification with one's natural environment; and secondly, a projection of some parental qualities onto the environment and an introjection of them into the psyche. As a result, identification with the lost environment gives a sense of identification with lost parental figures – the mountain becomes a father, the river becomes a mother, and our return to the natural environment feels like a homecoming. Naess gives a good example:

> the Lapps of arctic Norway have been hurt by interference with a river for the purpose of developing hydroelectricity. Accused of an illegal demonstration at the river, one Lapp said in court that the part of river in question was "part of himself".[3]

The Lapps' feelings demonstrate the process of identification and introjection through which the natural environment becomes "part of the self": the child exists in the external environment – feels the temperatures, breathes the air, experiences the rain. They run in the meadows, jump and swim in the rivers, climb the trees, and pick their fruits. The meadow, the river, and the trees indulge the child, satisfying their need for freedom, exercise, and play without demanding anything in return. The child feels emotionally safe and secure. Images of specific natural objects are introjected and become part of the self. Naess talks about strong identification, but without introjection, nothing can become the part of the self. When we grow up and need a holding environment, we go to nature if it is available. I am reminded of the Kindle logo, which shows a child reading from a Kindle while sitting under a tree large enough to provide shade and shelter. The child is alone and not afraid because they feel protected and safe.

One of my patients spent most of her time outside the home from the age of four or five.

Her parents often fought bitterly, and the child felt insecure and threatened. Being outside gave her a sense of relief. This feeling has remained with her even as an adult. Having returned from a trip two weeks earlier, she related in our session: "I enjoyed the trip. It had everything I liked – a river, a few hills, we also trekked a little, and we drove through dense forest on both sides of the road." Nature is her escape from unpleasant reality. Empirical research also supports this claim. Hundreds of studies in environmental psychology have demonstrated the positive effects of nature in the treatment of human physical and psychological illnesses. It is well known that the view of even a single tree from the window of a hospital bed significantly improves the prognosis. However, there is a lack of original research on the relationship between humans and nature from a psychoanalytic point of view.

In the research paper "The Human-Nature Experience: A Phenomenological-Psychoanalytical Perspective" (2018), Robert D. Schweitzer, Harriet Glab, and Eric Brymer conducted intensive interviews with nine participants about their lived experience of nature and its significance for their well-being. They conclude that natural objects are our internal objects and our self-objects. Nature has a holding function, like the mother. As Donald Winnicott pointed out, this holding function is extremely important for the survival of the infant. Not only does it meet all the physiological needs of the infant with its daily changes, but its most important function is to create a nurturing environment in which the infant also feels warm and psychologically secure. Winnicott goes so far as to say that the absence of this holding function is likely to have a serious effect on the infant's mental health. In the paper, two of the interviewees reflect:

> Well, you know, they call it Mother Nature. That's an appropriate term. It [natural world] is where I come from. So I am connecting back to myself by connecting with nature because I came from it.
>
> (Jen, 30 years old)

> I think people talk about the natural world as something completely separate to us, but we are nature as well and I think we just forget that. . . . Nature is like, "You're welcome." It always feels like home. It really is a return to.
>
> (Daisy, 27 years old)[4]

Similarly, in *Environmental Melancholia: Psychoanalytic Dimensions of Engagement* (2015), Renee Lertzman concludes – based on detailed qualitative research – that our claim of being apathetic towards nature is false. Rather, apparent apathy is a sign of mourning for the loss or destruction of nature. It is a defence. The lived experience of these subjects is similar to the one we have with our internalized parental images. External nature is therefore not merely external but also internal.

Nature is not always experienced as benign. Natural disasters such as floods, landslides, cyclones, storms, and tsunamis shake up this image of nature as a holding mother. In our trauma work with the victims of the 2004 tsunami, we visited a number of schools in the worst affected area of India. The children had decorated the blackboards and walls of their school with beautiful pictures of the sea and fish. After the tsunami, they felt threatened by the sea and drew pictures of the sea as a monster, with dead bodies lying on the beach. However, it is not clear whether these images will remain dominant. Perhaps, as with our parents and with the slow passage of time, we will identify with them as the most important, even if they evoke ambivalent feelings. Understanding how deeply we are connected to nature is vital to ensure its preservation. I believe that the best way to conserve nature is not to show how much I am sacrificing to conserve nature, but to realize through identification that in my attempts to preserve nature, I am conserving myself.

Notes

1 Arne Naess, quoted in Stephan Bodian, "Simple in Means, Rich in Ends: An Interview with Arne Naess", in *Deep Ecology for the Twenty-First Century*, ed. by George Sessions (Shambhala, 1995), pp. 26–36 (p. 26).
2 Arne Naess, "Self-Realization: An Ecological Approach to Being in the World", in *Deep Ecology for the Twenty-First Century*, pp. 225–39 (pp. 230–31).
3 Naess, "Self-Realization", p. 231.
4 Robert D. Schweitzer, Harriet Glab and Eric Brymer, "The Human–Nature Experience: A Phenomenological-Psychoanalytic Perspective", *Frontiers in Psychology*, 9 (2018), https://doi.org/10.3389/fpsyg.2018.00969, pp. 9 and 10 [accessed 27 August 2024].

15 Out of Paradise

The Future of an Ecological Disillusionment[1]

Luc Magnenat

The Illusion of a "Vast Dwelling With Two Superimposed Planes"

Our so-called Western culture, which Philippe Descola calls "naturalism",[2] is defined by the *discontinuity* of interiorities between humans and "other-than-human"[3] living beings and by the *continuity* of physicalities with other-than-humans. Humans alone are said to have a soul, reflexive consciousness, reason, and cultural diversity, while all beings are subject to the same universal laws of nature, sharing the same atomic, cellular, genetic constitution, and so on. This cultural ontology maintains an "illusion" of nature (shared by humans and non-humans) as universal and of culture (specific to human beings) as relative. Other-than-humans exist only according to the functions they perform for humans – as resources, servicers of the ecosystem, objects of scientific observation, or aesthetic contemplation. A divide thus organizes this form of culture, which, like "a vast dwelling with two superimposed planes", elevates human beings above nature by virtue of a culture and a supposed reason that other-than-human beings lack.[4]

This cultural ontology has proved extremely fertile at the scientific level, providing the bedrock for the industrial and technological revolutions. It is, however, also the bedrock of the environmental crisis. Today, this crisis is "disillusioning" us by inflicting a double loss: the ongoing, objective loss of the rich biodiversity and stable climate of the Holocene, and the subjective loss of our faith in the ecological viability of naturalist culture. The experience of this double loss is put into perspective in this chapter by exploring the way in which a totemistic culture – the North American Indian Crow people – went through an environmental and cultural disaster similar to the one we are being dragged into today: an environmental crisis that is also a social crisis. Using the Crow story as an example, I want to show how different the ecology of other cultures (animist and totemistic, for example) is from the ecology of our naturalistic culture. From this emerges the idea that the cultures of native peoples could anticipate the cultures of the peoples of the "end ages". These cultures could constitute our "memory of the

DOI: 10.4324/9781003634959-19

future" and support our faith in the future through the example of their capacity for resistance in historical contexts of environmental disasters and extreme social violence.[5]

The Biosphere as the Ego's Objective and Subjective "Security Background"

Humanity is no stranger to disaster and adversity. However, ecological research in the environmental sciences tells us that the environmental disaster towards which the Anthropocene is headed differs from other disasters that humanity has experienced. Previous catastrophes have taken place against a fundamentally stable *planetary* environmental background. With the current mass extinction of species and increasing climate disruption, we are losing the rich biodiversity and generally stable climate of the Holocene biosphere that has been the "backing" of our cultures and our personal subjectivities for 11,000 years. With the loss of this biosphere, I believe we are losing what James Grotstein calls a "background object of primary identification".[6] This is the object of support that a baby experiences when, sitting on his mother's lap and leaning against her, he explores the world of object relations (culture) before him, unconsciously leaning on the security of the symbiotic bond between his back and his mother's body (nature). This perspective, which stems from the clinical work on autism, suggests that our cultures not only influence us, externally and internally, in the manner of a paternal superego, but also that culture is rooted in the traces of our earliest sensory-motor learning.[7] Our cultures support us in the manner of a maternal backrest, while themselves being inescapably supported by the biosphere. Nature and culture thus form what we might conceive of as "a vast dwelling with two articulated and co-constructed planes", constituting the meta-frame that guarantees our thinking activity.[8] From this perspective, nature appears less as a world to be tamed (like the id) than as a "universal background", a "referent posited without discussion", as a reality that both resists and supports our ego.[9] In other words, the construction of this background object is not only intersubjective, akin to a sensory-motor and emotional attunement between mother and baby; it is also "transsubjective":[10] the corollary of a mother's link to her social, cultural, and environmental milieu, a connection occupying the "ambiguous position" described by José Bleger through the coalescence of our "syncretic egos".[11] As Sigmund Freud wrote, "individual psychology is also immediately and simultaneously a social psychology".[12]

Consequently, by losing the environmental support of the rich biosphere and stable climate of the Holocene, we lose what Joseph Sandler describes as a "background of subjective security of the ego", a feeling that Sandler relates to "the permanence of the presence of familiar things", that is, the permanence of our living surroundings.[13] Today, we are collectively walking away from this "Paradise", as the title of this chapter suggests. The frameworks of our lives are changing drastically, and we have to learn to live with what Yolanda Gampel calls a "background of disquieting subjective strangeness of the ego".[14] The environmental crisis introduces a caesura between our experience of yesterday's world and our anticipation of tomorrow's, a caesura that renders obsolete our present understanding of a world that is becoming strange and foreign, and which is no longer temporally reliable.

I believe that the emergence of this background of disquieting strangeness is traumatizing, impacting our patients and our fellow citizens alike, and is experienced as doldrums or as feelings of distress, a loss of meaning in our work, or a sense of futility about our lives. Along with Silvia Amati Sas and René Kaës, we can attribute these feelings of disquieting strangeness to the preconscious awareness of a flaw in the containment offered by the structures of our lives.

The Human Being Is Just One Stitch in the Fabric of Life

In 1917, Freud noted the humiliating effects of science on human self-esteem. After Copernicus and Galileo discovered that planet Earth is not the centre of the universe, Darwin acknowledged that the human species is part of the evolution of all other species, and Freud himself declared that the human being is directed by his unconscious and inhabited by a death drive, it is now the discoveries of the environmental sciences that are inflicting a new "narcissistic wound" on mankind.[15] The notion of the ecosystem implies that the human being is but a stitch in the fabric of life, a "child of the biosphere", as dependent on this ecosystem of ecosystems as an infant is on its parents.[16] Moreover, the environmental sciences reveal that humanity is undergoing an anthropogenic environmental crisis that is beyond the human capacity to control it. An ecological humiliation is thus added to the cosmological, biological, and psychological humiliations listed by Freud.

The notion of the ecosystem greatly enhances our feelings of insignificance and vulnerability by heightening awareness of our interdependence with other living species and potentially extending what we might consider our "community of belonging" to the other-than-human world of microorganisms, flora, fauna, and our waste. Ecologists, those clinicians of the biosphere, are thus the agents of disillusionment: with ecological ecosystem thinking, what is out of sight is no longer out of our world, and nor should it be out of our thinking, because the world of "elsewhere" that we fantasize our waste going to no longer exists.[17] In the ecosystemic world of the biosphere, with neither inside nor outside, what we evacuate comes back to us in the form of an environmental crisis composed of the accumulated waste of three centuries of industrial and technological revolutions.

As a result, climate disruption and the ongoing erosion of biodiversity call into question the *disjunction* between the history of nature and the history of humankind, one colliding with the other: microorganisms, flora, fauna, and our waste have now entered our psychoanalytical clinic, our political life, and even, in the case of our waste, every drop of rain, every cell in our bodies.

Environmental science teaches us that *Homo sapiens* is not only a great maker of tools, objects, and waste; he is also, unwittingly, a maker of hyperobjects.[18] A hyperobject is an object that is immeasurably extended in time and space. Galaxies, black holes and the biosphere are all hyperobjects, just as a mother's womb is for the foetus. The sum of the chemical and nuclear waste discharged on to our planet and the machinery of the industrial and technological revolutions are all hyperobjects. The environmental crisis is a hyperobject made up of the sum of the all-too-real consequences of the all-too-human development of technologies, and its discovery by the environmental sciences radically belies the human fantasy of being the "lord of this world".[19] We have been consumed by the environmental crisis like Jonah in the belly of the whale: the second disillusionment inflicted on humanity by ecology! This, I believe, is our new existential background of disquieting strangeness.

Is it therefore conceivable that the "end of the world" – or at least the end of our representation of the world – has already occurred? If there are actions to be taken, commitments to be made, they are *not*, it seems, in the sense of preserving what the hyperobject has already caused to disappear, nor of making the hyperobject itself disappear. The latter has subsumed us in its ontology. It is thus conceivable that a temporal disruption is underway which means that our *future* is perhaps already *present* – included in the ontology of a hyperobject formed in the *past*.

Freud may have seen man as a "prosthetic god" – a being who could maintain the illusion of feeling divine through the powerful techniques that improve his life and the functioning of his organism.[20] Today, however, the forever neotenic being that we are seems to be crushed, rather than elevated to godlike status, by a "prosthesis" of technical objects and waste. Their cumulative presence imposes itself on us in a brutally *interobjective* way in an environmental crisis because our waste is a "zombie" whose lifespan considerably exceeds not only the duration of its use but also the duration of a human life. We could say that waste spends its long life becoming a hyperobject and constituting the environmental crisis; it becomes those climatic "tipping points" that we now perceive to be packaging climate disruption through positive feedback mechanisms – beyond any measures humanity as a whole might take.

The "Mother Idea" of Our Age: "Dreaming" the Hyperobject We Have Created

As a psychoanalyst, I can see emerging in this hyperobject what I will call a "mother idea".[21] The notion of the "mother idea" appears only a few times in Wilfred Bion's work as a term used to define the emergence of a "collective new idea". It aims to designate the proto-mental action of beta-elements coming from what Bion calls "the infrastructure of the real". The mother idea, as an aggregate of beta-elements, is therefore composed of "things-in-themselves, feelings of depression-persecution, aspects of the personality linked by a sense of catastrophe", says Bion.

This is why the mother idea of an epoch is a source of raw emotions whose actions exert pressure as an "over-generator" of individual and collective thoughts, notes Francesca Bion.[22]

In abstract Bionian terms, the mother idea is an emergence from O, the field of proto-mental formlessness – into K, the field of knowledge through the process of symbolization. The mother idea is what Thomas Ogden might describe as a "nocturnal terror", or a proto-mental unthought waiting to be "dreamed", as one of the elements of our *Hilflosigkeit* [helplessness] in Freud's terms, our original, fundamental distress. The mother idea of an epoch constitutes a driving force behind the work of culture in that epoch. The emotional nature of this driving force is "a feeling of catastrophe" (β) waiting to be heard as a "sign of catastrophe" (α). This occurs through the development of an alpha function endowed with a "sense of catastrophe", which roughly means it is endowed with an ability to transform a feeling of catastrophe by putting it into words.[23] This, I believe, is the only way out of the distress of discovering the environmental crisis. By thinking about the mother idea of our time, we should be able to collectively improve our actions – to try not to act against each other, but above all – and I want to emphasize this – to find faith in the future: faith that a sense of catastrophe can be acknowledged; faith that others are gifted with the reverie of emotional life; faith that others are not "obstructive";[24] faith that a simple, emotional, living encounter with others can transform a feeling of catastrophe into a catastrophe that makes sense.

The "Limit-Experience" of the Crow People

Meltzer asked himself a question regarding certain patients: "what is the emotional experience, the undreamt dream, that these patients can not yet dream and that we should manage to dream for them by allowing ourselves to be carried along by a capacity for reverie?".[25] I postulate that the progressive bursting of the bubble of collective denial of the environmental crisis – and its consequence, our contemporary Sphinx-like doldrums – poses a similar question to us: "What is the undreamt dream that our world must now attempt to dream?"

I believe that the way forward in this enigma is to employ to the point of view of a third-party, an object endowed with reverie. I search for this gaze among those peoples whom Freud considered "primitive", but whose cultures are shaped by experiences "at the limits": by extreme geographical and climatic conditions, and with potentially crushing contexts of social injustice and political oppression. I believe that the cultures of these peoples, and the richness of their capacities for resistance, are potential sources of faith in the future for us today.

We could consider the Tuareg and the Inuit peoples,[26] but I propose instead to look at a North American Indian people: the Crow. The Crow were nomadic hunter-gatherers and warriors whose way of life depended mainly on buffalo, which were slaughtered en masse by European settlers in the nineteenth century. With the disappearance of the buffalo, the Crow people lost the background object of their culture that nature had provided. In his 2006 book *Radical Hope*, Jonathan Lear paints a picture of this Indigenous people and their customs. At the heart of his account is the environmental, cultural, and social disaster that the Crow experienced when European settlers invaded the American West. This catastrophe raises the question: What happens to a subject and to a society when the means of living according to traditional ideals collapse, when the lifelong task of becoming a hunter and warrior in order to lead a good life becomes impossible?

This is what the Crow people experienced when their culture was attacked and their ideals became impossible to achieve. While our cultural fracture will be of a different quality from that faced by the Crow people as targets of imperialism, this is a form of cultural fracture that each of us can experience today. However, the Crow people, as "political animals",[27] did not adapt to it – which is precisely what allowed them to endure, even at the risk of losing their identity. It is in recognizing this point that Lear's account becomes particularly interesting. Lear describes how the Crow came to terms with the devastation of their culture by finding a compromise between their traditional values and those of the emerging new world by managing to maintain a sense of "continuity of being" in the midst of catastrophic change. The transmission of Crow culture was oral. Lear's work draws on the rare testimony of one of the last great Crow chiefs, Plenty Coups, because of his many acts of bravery. The fact that he wanted his account of Crow history to be recorded in writing was a sign of his awareness of the *value* of knowing how his people survived, his awareness of being an individual representative of the collective transformation that took place, and his hope that a future generation would be able to make his story their own.

Lear quotes Plenty Coups's comments on his experience of the Crow disaster which saw Crow culture devastated by famine, malnutrition, disease, defeat, and confinement to a reservation: "When the buffalo went away, the hearts of my people fell to the ground, and they could not lift them up again. *After that, nothing happened.*" Plenty Coups seems to imply that when a familiar world breaks down and life becomes strange, the familiar good things of life can no longer happen. One is uprooted, threatened by the prospect of a ghostly, ambiguous existence.

The activities characteristic of a "good life" cease to be intelligible in a radically new and threatening historical and cultural context. As Eigen points out,[28] it is at such a point that culture may be unable to hear or bear the alarm signals of impending catastrophe. This seems highly relevant today. When a culture is challenged to dream a new, undreamt dream, it has to endure a life that has become incomprehensible, waiting to develop new identificatory projects.

As a psychoanalyst, I can hear in this statement the threat of being swept away by a melancholic identification with both a culture and the surrounding natural world being felt as a lost mother or background environment. Above all, I hear in it an aptitude for suffering, a feeling of catastrophe, a capacity to register rather than turning a blind eye – in short, for endurance as a mobilizing sign of catastrophe, apt to develop a "sense" of catastrophe. As an analyst, I therefore feel at odds with Plenty Coups: I think it is wrong to say that "nothing happened after that" because, importantly, Plenty Coups subsequently listened to his tribe's feelings regarding the catastrophe and was able to interpret them.

A "Tribal Dream"

The Crow discovered a courageous way of facing up to the collapse of their world (both the natural world and culture), and this forms the core of Lear's work. One Crow tradition was to encourage pre-pubescent youth to go into seclusion in the wilderness to have a dreamlike vision, which was then told to tribal elders who interpreted its meaning. In Crow culture, humans have a meaningful place in the more-than-human world that includes soils, skies, flora, fauna and spirits. When aspects of these worlds go wrong, the Crow feel wrong, and they look for a visit from a spirit in a dream to explain how the world might be readjusted, or how they might adjust to a different world. The conceit of a spirit's visit in a dream could be likened to the psychoanalytic notion of "narrative derivatives" of the unconscious dream field co-created during the analytic encounter.[29] Plenty Coups experienced this initiatory rite of passage at the age of nine in the context of his tribe's anxiety about the future: this is the waking residue. He reported a very long dream, from which I quote two extracts that were particularly significant for the members of his tribe:

> Out of the hole in the ground came the buffalo, bulls, cows and calves without number. They spread wide and blackened the plains. Everywhere I looked, great herds of buffalo were going in every direction, and still others without number were pouring out of the hole to travel on the wide plains. When at last they ceased coming out of the hole in the ground, all were gone, all! There was not one in sight anywhere, even out on the plains. I saw a few antelope on a hillside, but no buffalo [. . .]. Then out of the hole in the ground came these bulls, cows and calves past counting. These, like the others, scattered and spread on the plains. But they stopped in small bands and began

to eat the grass. Many lay down, not as a buffalo does, but differently, and many were spotted. Hardly any were alike in color or size.

[. . .]

"Listen, Plenty Coups," said a voice. "In that tree is the lodge of the Chickadee. He is lest in strength but strongest of mind among his kind. He is willing to work for wisdom. The Chickadee-person is a good listener. Nothing escapes his ears, which he has sharpened by constant use. [. . .] He never intrudes, never speaks in company of strangers, and yet never misses a chance to learn from others. [. . .] The lodges of countless bird people were in the forest when the Four Winds attacked it. Only one remained unharmed, the nest of the Chickadee. Develop your body, but do not neglect your mind, Plenty Coups."[30]

Yellow Bear, the wisest of the group of elders, offered this interpretation:

He has been told that in his lifetime the buffalo will go away forever and that in their place on the plains will come the bulls and cows and calves of the white man. I have myself seen these spotted buffalo drawing loads of the white man's goods. And once, at the big fort. [. . .] I saw cows and calves of the same tribe as the bulls that drew the loads.

Plenty Coups' dream means that the white man will conquer this country and hold it, and that their spotted buffalo will cover the plains. He is told to think for himself, to listen, and to learn to avoid disaster by listening to the experience of others. He is advised to develop his body without forgetting his mind. The meaning of this dream is clear to me. I see it as a warning. The tribes that fought the white man were all defeated, wiped out. By learning like the Chickadee, we can escape this and keep our land.[31]

Lear contextualizes this dream and its interpretation by clarifying two points from within the Crow tradition. The "Four Winds attacking the forest" represent the four dominant winds known to the Crow, four deities of their totemic world. In the dream, these Four Winds are unleashed simultaneously, which never happens in reality; they could symbolize a chaotic disaster, an upheaval of world order. These Four Winds of chaos attack a forest which, in the interpretation, represents all the Indian cultures settled in the plains and mountains of the American West. They are, I believe, the dreamlike figuration of distress, a feeling of annihilation, and their status as dream characters gives them the value of a sign of catastrophe. Before this sign, the tribe is in desolation, in literal catastrophe; after this sign, the disaster takes on meaning and the tribe reaches a solidarity that binds around a new dream character.

Birth of a Totem

The Chickadee is not just an undistinguished other-than-human inhabitant of the Crow environment. It is a bird whose habits the Crow know very well. They know that silence and listening are qualities of this bird that distinguish it from many other species. The Chickadee is both a mythical figure in their culture and an unconscious symbol of the workings of their minds.

Yellow Bear's interpretation of Plenty Coups' dream is an expression of the depressive position, with its integration of friends and enemies, past and future, within an emotional framework of mourning. The Chickadee emerges as a "selected fact"[32] that is both the content of tribal reverie and the figuration of a *background of subjective security of the ego* that is regenerated, both ethical and containing. The whole tribal dream scene appears as a working group at work, this group function whose characteristics, says Bion, are analogous to those Freud attributed to the ego, and to those of a mother endowed with reverie and holding. The tribe's weaving of thoughts brings hope to the heart of catastrophe.[33] It creates a *totem*, a deity, the Chickadee.

Contrary to what Freud suggests in *The Future of Illusion* (1927), this religion is *not* an illusion for the Crow: this totem helps them live through the fallout from the paradise of their traditional culture in order to better anchor themselves in a new historical reality.

The Crow have demonstrated a keen sense of the falling out of the world, along with its corollary: a faith in their ability to find a way through this existential impasse. Through this dream and its interpretation, by identifying with the Chickadee, the Crow were able to face up to a radically new historical reality: the onslaught of the white man's incursions and violent takeover along with the disappearance of the buffalo. The Crow used this dream as a source of ethical guidance which inspired their political relationship with the American government. They suffered less than other Native American tribes from the ethnocidal actions of the so-called conquest of the West and were able to preserve some of their lands.

The Modest Omnipotence of an "Object to Save"

Lear notes how the tribal dream was a source of radical hope for the Crow in the past and might be for us in the future: the fundamental hope of a rebirth, a return to a life in a form not yet intelligible at the time of the disaster. In this sense, the Chickadee proposes a collective identificatory project, an "ideal of we", as Norbert Elias would put it.[34] This hope seems to have sustained the Crow people's courage to find a third way between two extremes: between the suicidal excess of the temptation to wage war against the American government and the bankruptcy of regressively losing oneself to the ambiguity of "adapting to anything whatsoever" even to a totally alien culture.[35] Although they were able to integrate into European-American culture, the Crow remained Chickadee.

The Chickadee appears in this story as an expression of ego function, both that of listening and of the negative capacities of the ego open to learning about the emotional life of the world.[36] The Crow have not stifled their sense of catastrophe in a regression to the ambiguity of the Four Winds, but they have created a totem that reflects the development of their self-preserving capacities in the face of social and environmental catastrophe. Their unconscious has not been oblivious to history, but on the contrary, it has profoundly been shaped by the dreamlike mentalizations induced by their historical context. This, I believe, is how the Crow dreamed the undreamed dream that their totemic culture was forced to dream under the pressure of its encounter with naturalist culture. They gave themselves what Amati Sas evocatively refers to as an "object to save": the Chickadee. For Amati Sas, the notion of an "object to save" conveys a concern for the existence, destiny, integrity, and dignity of another subject. She states that the "object to save" is nothing more than the good internal object with which we are all familiar. As an example, she recalls a woman who had been tortured by the henchmen of a South American dictatorship and who, during her psychoanalysis, said: "I had to resist, because I couldn't accept that my child could live in a world run by these people." For this patient, this thought had been a pre-conscious human thread that sustained her sense of existence, even at the height of a highly traumatic situation.[37]

As a psychoanalytic concept, the "object to save" preserves the idea of the human as opposed to being treated as a dehumanized, massified thing, notes Amati Sas. The remarkable thing about the ability to imagine an object to save is that this thought aspires to save others, such as a child in the example given earlier, in both the external and internal worlds. At the same time, it preserves the thinking ability of the person under duress who is able to imagine an object to save. The constitution of an object to save intrinsically implies the modest power of the healthiest part of our personalities, the capacity to preserve a link with our own minds and between self and other. Amati Sas points out that to have an object to save is to care for another human – indeed, for an "other-than-human", a soil, a sky, a flora, a fauna – in a way that involves a form of magic, a kind of "modest omnipotence" analogous to that of a mother caring for a child: the magic, I think, of generating a thought for another despite being absorbed by the social violence of a catastrophic historical context.[38]

Will the Cultures of the First Peoples Become the Cultures of the Last Peoples?

In the aftermath of history, could the Crow become the repository of our contemporary cultural identificatory project? Could the cultures of the first peoples anticipate the cultures of the last peoples, the cultures of those who, in all likelihood, will have to live in increasingly extreme climatic conditions in the future?

Descola's *L'écologie des autres* [*The Ecology of Others*] (2011) is a telling title. It refers to the ecology of cultures other than Western naturalism, in particular animism and totemism. When an animist Amazonian Amerindian does a slash-and-burn, gathers a plant, or kills an animal for food, as psychoanalysts, we might think that he is acting with the unconscious feeling of committing incest or cannibalistic murder. Rituals of atonement and respect for a nature felt to be human follow. And we have seen just how much the Crow, who are totemic, respect other-than-human species – think of the bison and the chickadee – as part of their own subjectivities.

Consequently, animistic and totemist cultures approach the natural world with respect. A mutually fertile bond is established, whereas we Westerners are culturally inclined and doomed to be mere observers, predators, and depredators. Cultures such as the Crow inhabited the American West for millennia before European naturalists invaded and appropriated their lands. We could say that the Native American peoples dreamed the Great Plains. The fertility and biodiversity of the American West dazzled the first European settlers. The biodiversity of these plains was the legacy of the totemistic culture of the Indigenous peoples before it was desertified by the plundering of its resources by a rampaging, naturalistic culture.

The challenge posed by the mother idea of our times, the discovery of the environmental crisis, is either to disappear with an exhausted form of humanism or to undergo a cultural metamorphosis by opening up to other cultural ontologies: experimenting with diplomatic relations between humans and other-than-humans that respect our place on this planet: that of children of the biosphere, mere stitches in the fabric of the living. Descola sets this out:

> It's up to each of us, wherever we find ourselves, to invent and nurture modes of conciliation and types of pressure capable of leading to a new universality, at once open to all the components of the world and respectful of some of their particularisms, in the hope of warding off the distant deadline when, with the extinction of our species, the price of passivity would be paid in another way: by abandoning to the cosmos a nature that has been orphaned by the failure of its authors to concede it genuine means of expression.[39]

This, I believe, is how Descola dreams the undreamed dream that our naturalist culture is forced to dream under the pressure of the mother idea of our time. Descola offers us ethical guidance. If we are to accept that hatred, omnipotence and destructive narcissism are part of our internal world and inscribed in the ontology of our species – that uncare thrives when it is not overridden by concern for life, as Sally Weintrobe points out – then it is necessary that the non-psychotic, benevolent part of our personality, that which "takes care" of the human and other-than-human, works constantly to *contain* this psychotic functioning. What Weintrobe calls the Exception (a form of our destructive narcissism) must be only part of our psychic life, but *must not rule* it; otherwise the hubris of deregulatory madness sets in, both internally and externally. Weintrobe examines what happens when the healthy, adult part of our personalities and societies loses the power to contain the Exception's uncare, resulting in a culture born of collusion between neoliberal economic ideology and the psychotic part of our personalities and societies.[40]

An Autumn Landscape in the Northern Forests

In a naturalistic culture, remaining human in the face of environmental collapse might take the form of developing an alpha function of the ego better equipped for ecosystemic thinking, in the manner of the Crow. Aldo Leopold, the father of environmental ethics, evokes this better than I can by the sheer force of a writing style that speaks to the reader's intuition:

> An autumn landscape in the northern forests is the land, plus a red maple, plus a crested grouse. In terms of conventional physics, the grouse represents only one millionth of a hectare, in terms of both mass and energy. Yet if you subtract the grouse, there's nothing living left in the bargain. An enormous quantity of something like motive power has been lost along the way.[41]

From Bion's perspective, the development of thought oscillates between a normal schizo-paranoid position and the depressive position. This oscillation is modulated by the emergence of what Bion calls a "selected fact". For the analyst, this translates into listening to the patient's apparently disparate utterances and behaviours along with his own potentially chaotic counter-transference. At some point, a kind of Eureka moment emerges that gathers and binds these scattered, schizo-paranoid analytic bits into a selected fact. This can usher in a different experience, located more in the depressive position, with a fuller, more realistic, and symbolic representation of the emotional situation of the session. I think that Leopold's stylistic use of the "crested grouse" in this quotation gives us an idea of such a selected fact – that enables us to intuitively "grasp" the nature of an ecosystem.

Leopold is a distant descendant of the European settlers and naturalists who invaded the American West. While he may feel a kinship with nature – sensing its beauty even in its most modest forms, such as a marsh, a scree, or the shape of a river's orbs – and while he may understand its ecosystemic workings profoundly and scientifically, the cultural ontology of his sensibility is radically different from that of Plenty Coups. For the (naturalistic) culture of the one is *not* the (totemist) culture of the other.

Consequently, the nature of one is *not* the nature of the other, and the ecology of one is *not* the ecology of the other.[42] Yet, like the dream of Plenty Coups, Leopold's affectionate and poetic writing seems to speak an *originary* language: that of the beauty of nature, that of the catastrophic sense that an ecosystem could disappear, that of projective and adhesive identification with natural environments, of animals perhaps.[43] His writing makes us *dream* in a way that unknowingly develops a function of the reader's ego: an ability to intuitively feel the beauty of an ecosystem in operation – an autumn landscape in the northern forests – *and* to sense the catastrophe that this ecosystem could disappear with the disappearance, merely suggested by Leopold, of a crested grouse.

I believe that a voice such as Leopold's, among many other ecologists, speaks to us of our contemporary "object to save": the ecosystems of the more-than-human world that environmental sciences are introducing us to. Leopold's writing, as I feel it, transforms our sense of catastrophe into faith in the future and contributes towards dreaming the fundamentally undreamt dream of our naturalist culture by weaving a new ethic: "Progress is not to bloom roads in landscapes that are already wonderful, but to bloom receptivity in human brains that are not yet so."[44] I do not regard this as a matter of defending or protecting nature but, like Leopold, of being "nature defending itself"[45] by letting ourselves be socialized by it, suffering with it, enduring, and sharing our feelings of catastrophe. This, I believe, is how Leopold dreams the undreamt dream that our naturalist culture is now challenged to dream: he identifies with nature; he does not dominate it.

What Alarm Signals for Our Feelings of Catastrophe?

Psychoanalysis can no longer afford to ignore the new environmental and social reality. What social geography should we refer to in order to find our place? What emotional compass will orientate us to this unknown? What are the signals for our protomental feelings of alarm? In *Origins of Totalitarianism* (1951), Hannah Arendt describes the unprecedented originality of a totalitarian organization that makes the members of society act according to the rules of a fictitious world, a delusional ideology, the expression of radical evil. This radical evil has no history, she writes. Rather, in evoking the origins of totalitarianism, Arendt describes a constellation of elements always more or less intensely present in every society: anti-Semitism, the corruption of the nation-state, racism, imperialism, and the alliance between capitalism and the masses, for example. To these elements of the banality of evil, I think we could today add others: those of the "banalization" of the environmental crisis, of the nuclear threat, of social suffering and injustice, of lying on social networks and in the mass media, and, last but not least, the banalization of delegating the activity of thinking to artificial intelligence.

The important point for Arendt is that the presence of these elements can intensify until they crystallize and emerge in the historical form of a totalitarian regime, of what I would call a "negative selected fact". The dynamics of the destructive individual and collective impasses described by Weintrobe could correspond, from another perspective, to what Arendt describes as a crystallization of the elements of totalitarianism.[46] This, I believe, is how Arendt dreams the "undreamed dream" of our political life, informing us of the metahistorical, transubjective roots of the life drive and the death drive. As a counterpoint, in *Human Condition* (1958), Arendt locates the possibility of a non-totalitarian world of resistance and rebirth contained in the human capacity to construct and preserve a political sphere, a space for co-thinking. She writes: "The miracle that saves the world, the realm of human affairs, from normal, 'natural' ruin, is ultimately the fact of natality, in which the faculty of acting originates ontologically."[47] From a psychoanalytical perspective, the fact of natality links with procreation, to which we owe life, and to the Oedipal situation, to which we owe finding a place in this life, within a biotic community. We must add to this the emotional encounter with others by which thought as an activity and a political space are constructed. We are once again seeing the elements of totalitarianism coming together in our contemporary world, crystallizing into new tyrannies and new wars. I hypothesize that these elements of totalitarianism are the index, the sign, of true catastrophe, the alarm signal of our collective distress at the emergence of what I call the mother idea of our age.

Harold Searles and Plenty Coups

The role of the ecologist and the psychoanalyst is to reveal to others what they don't want to know. They are forced to say what no one wants to hear, that is, to inform people about a humanity governed by its unconscious and by a hyperobject – the environmental crisis – born of its unconsciousness. Ecology and psychoanalysis can only be marginal presences on the world map. Insofar as they enunciate unpleasant scientific truths that disillusion us by illuminating the fantasy of the human being as the lord of the world, they convey the discontent in culture, the dissatisfaction with culture, and can only potentially be enemies of one and all. This is what makes them our objects to save, our Chickadees, within the culture of the limit-experience of the game with death in which it falls to us to live.

As I conclude, I can only think back to Harold Searles' intuition:

[B]ecause we tend to feel that sudden death from nuclear warfare is a threat entirely out of our control, we may prefer the slower, more controllable death that pollution [the environmental crisis] offers as seemingly the only alternative. We know that pollution is a process to which we contribute daily; it is something which, in however small part, we know we actively do. On the other hand, to regard such slow strangulation as an inevitable agony is to yearn for the quick relief that nuclear warfare would bring.[48]

(my parenthesis)

With these words, Searles appears to acknowledge that we are collectively threatened by the fantasy of an "apocalypse without a kingdom". I borrow this term from Günther Anders, for whom the Christian myth promised us an apocalypse *with* a kingdom – with the possibility of salvation through redemption, through the survival of our works and through symbolic survival – whereas the great hopes of the Enlightenment promised the kingdom of progress *without* the apocalypse.[49] Since Hiroshima, with the ongoing nuclear threat and the experience of a runaway environmental crisis, we have been living in expectation of an apocalypse without a kingdom; in other words, of absolute destruction that no salvation can redeem. What Hiroshima annihilated, Anders adds, was History itself, by ushering in a new temporality, that of the "delay" that separates us from a truly absolute end, the "end of time" for humanity. As long as we live in a delay, in the 'time of the end' before the "end of time", we no longer have to ask ourselves how long the delay will last. Only one question remains, notes Jean-Paul Engélibert: "How can we escape it? How can we disprove the prediction and, from within the deadline, in the time we are given, resurrect a history?"[50] Identifying the atomic bomb and the environmental crisis as absolute threats, as Searles does, implies an absolute obligation: to do everything possible to ensure that neither history nor human thought disappears into nothingness. More precisely, I think that the mother idea of our time compels us to develop the *negative* capacities of our egos like the Crow, to become like "Chickadees" tuned in to the world, to our profound vulnerability, to that of the other species with which we cohabit, and to our intense destructiveness. It is a question, I think, of trying to live without memory, nor desire, nor understanding, but with a keen sense of catastrophe and faith in the future, to establish within ourselves the preconception of a history and a thought waiting to occur. This technical adage, and the attitude to life it implies, is, perhaps, our "object to save" as psychoanalysts, and an identifying project to be shared with others. For I believe that only by listening to our deepest anxieties – as Searles does – as manifested in external reality by the gathering together of the elements of totalitarianism or the elements of the natality, can we lay the foundations for an ethic adapted to the times in which we live.

In terms of psychoanalytical interest in ecology, I feel that Searles' 1972 article is something of an initiatory dream, like that of Plenty Coups. The violence of the existential dilemma before which this text places the reader – the slow collective strangulation by an environmental crisis or the end of time in a brief atomic flash – is thought-provoking. This nightmarish dilemma provides the material for a dream. The two terms of the existential vise described by Searles may seem equivalent in their reference to an "end of time" for humanity. But in fact they differ radically. One allows us to experience the delay of the "end times" emotionally, to dream it, while the other dispenses with it.

If I think of this dilemma "without memory, nor desire, nor understanding, but with a keen sense of catastrophe and faith in the future", an ethic takes shape. Nuclear war is the "apocalypse without a tomorrow" that must be avoided at all costs because a nuclear war and an atomic winter would not only constitute an immeasurable crime against humanity; they would also be radically biocidal.[51] They could wipe out a huge part of the earth's biodiversity along with mankind. On the other hand, an uncontrolled environmental crisis nonetheless implies the possibility of preserving a sense of humanity, sustained by the internal bearing of our "objects to save". It is compatible with "symbolic survival".[52] It is, I believe, an "apocalypse with a tomorrow", even if this "tomorrow" is made up of shreds of humanity surviving according to cultures close to those of the indigenous peoples we know, such as the Crow, within a biodiversity infinitely richer than the reduced biodiversity abandoned by an atomic winter. This biodiversity would carry a richer genetic memory of the biosphere's past and would therefore be a more fertile cradle for the evolutionary future of the biosphere of tomorrow.

Notes

1 This chapter was partly translated from the French by Andrew Weller, Paris, and reviewed by Donald Nicholson-Smith and Lynne Zeavin, New York. All translations of the original texts are by Andrew Weller.
2 Philippe Descola, *Par-delà nature et culture* (Editions Gallimard, 2005), pp. 241–79.
3 I adopt the term "other-than-human species" from Wendy Hollway, Paul Hoggett, Chris Robertson, and Sally Weintrobe (2022). Unlike the expression "non-human species", in my opinion "other-than-human species" better expresses the separation and differentiation of species *and* the fact that humans and other-than-humans belong to the same biotic ecosystem community, the biosphere. In other words, they share a relative universality of the biological, and even emotional and cognitive, foundations of life, while respecting the specific qualities of each species and ecosystem. To speak of "non-human" species is to focus too anthropocentrically on the "human". With "other-than-human" species, we live in a "hyperobject", the biosphere, in a "*more*-than-human world" (Morton, 2013). In other words, we share same "milieu" (an ecocentric term) as non-human species, rather than an "environment" (an anthropocentric term that situates human beings in the "middle" of the environment rather than as part of a set of ecosystems).
4 Descola, *Par-delà nature et culture*, pp. 91–127.
5 I refer here to Bion's psychoanalytical novel *Memory of the Time to Come* (1991), Larmor-Plage, Editions du Hublot, 2010, whose first part, entitled "The Dream", describes a post-apocalyptic world.
6 See James Grotstein, "Primal Splitting, the Background Object of Primary Identification, and Other Self-objects", in *Splitting and Projective Identification* (Jason Aronson, 1918), pp. 77–89.
7 See E. Bick, "The Experience of the Skin in Object-Relations", *International Journal of Psychoanalysis*, 49 (1968), pp. 485–86; D. Meltzer, *Explorations dans le monde*

de l'autisme (1975) (Payot, 1980); D. Anzieu, *Le "Moi-peau"* (Éditions Dunod, 1985); F. Tustin, *Autistic Barriers in Nevrotic Patients* (Karnc Books, 1986); and G. Haag, "Sexualité orale et moi corporel", *Topique* 2 (2004), pp. 23–45.

8   R. Kaës, *Le malêtre* (Éditions Dunod, 2012); and R. Kaës, (2020). "Notes sur les espaces de la réalité psychique et le malêtre en temps de pandémie", *Revue Belge de Psychanalyse*, 77, pp. 170–87.

9   Philippe Descola and A. Pignocchi, *Ethnographie des mondes à venir* (Éditions du Seuil, coll. Anthropocène, 2022).

10 I. Berenstein and J. Puget, *Lo vincular* (Buenos Aires: éditions Paidós, 1997). *Psychanalyse du lien.* Éditions érès, 2008.

11 José Bleger, *Symbiose et ambiguïté* (1967) (PUF, 1981).

12 Sigmund Freud, *Malaise dans la culture* (1930) (PUF, Quadrige, 1995).

13 Joseph Sandler, "The Background of Safety", *International Journal of Psychoanalysis*, 41 (1960), pp. 352–56.

14 Yolanda Gampel, "The Pain in the Social", *International Journal of Psychoanalysis*, 101.6 (2020), pp. 1219–235; and "Un mur tombe pendant la séance", *Revue Française de Psychanalyse*, 5 (2021), pp. 1253–59 (Paris: PUF).

15 Sigmund Freud, "Une difficulté de la psychanalyse", in *Œuvres complètes*, vol. XV (Paris: PUF, 1917), pp. 41–52.

16 See Luc Magnenat, *La crise environnementale sur le divan* (Edition In Press, 2019), and " 'Think Like a Mountain' – 'to Think of Oedipus': A Psychoanalytic Contribution to Environmental Ethics", *International Journal of Psychoanalysis*, 4 (2021), pp. 734–54.

17 Timothy Morton, *Hyperobjects. Philosophy and Ecology After the End of the World* (University of Minnesota Press, 2013).

18 See Morton, *Hyperobjects. Philosophy and Ecology after the End of the World.*

19 See Freud, "Une difficulté de la psychanalyse".

20 See Freud, *Malaise dans la culture* (1930).

21 Wilfred Bion, *Un mémoire du temps à venir* (1991). Lexique établi par Francesca Bion (Editions du Hublot, 2010), pp. 196–97, 208–9, 271–72, 274–75.

22 Bion, *Un mémoire du temps à venir*, p. 555.

23 See M. Eigen, "Towards Bion's Starting Point: Between Catastrophe and Faith". *The International Journal of Psychoanalysis*, 66, pp. 321–30, 325.

24 Wilfred Bion, *L'arrogance* (1957), in *Réflexion faite* (PUF, 1983), pp. 97–104.

25 D. Meltzer, *Explorations dans le monde de l'autisme* (Paris: Payot, 1975/1980).

26 See, for example, J. Malaurie, *Hoggar, Touaregs, derniers seigneurs* (Éditions Nathan, 1954); and *Les derniers rois de Thulé. Avec les Esquimaux polaires, face à leur destin* (Collection Terre Humaine, Éditions Plon, 1989).

27 See Wilfred Bion, *Recherches sur les petits groups* (PUF, 1961).

28 See Eigen, "Towards Bion's Starting Point: between Catastrophe and Faith".

29 See A. Ferro, *La psychanalyse comme œuvre ouverte* (Éditions érès, 1997), and G. Citvitarese and A. Ferro, *A Short Introduction to Psychoanalysis. The Psychoanalytic Field Theory* (Routledge, 2020).

30 Jonathan Lear, *Radical Hope. Ethics in the Face of Cultural Devastation* (Harvard University Press, 2006), pp. 69–71.

31 Lear, *Radical Hope*, p. 72.

32 See Wilfred Bion, *Aux sources de l'expérience* (1962) (PUF, 1979).

33 See J. Norman and B. Salomonsson, " 'Weaving Thoughts'. A Method for Presenting and Commenting Psychoanalytic Case Material in a Peer Group", *International Journal of Psychoanalysis*, 86 (2005), pp. 1281–98.

34 Norbert Elias, *La société des individus* (1987) (Éditions Fayard, 1991).

35 Amati Sas, "L'ambiguïté comme défense dans les situations extrêmes".

36 See Wilfred Bion, *L'attention et l'interprétation* (1970) (Payot, 1974).

37 See Amati Sas, "L'ambiguïté comme défense dans les situations extrêmes".
38 Amati Sas, "La modesta omnipotencia", *Revista de Psicoanálisis*, 5 (1997), pp. 21–31.
39 Descola, *Par-delà nature et culture*, p. 552.
40 See Sally Weintrobe, *Psychological Roots of the Climate Crisis: Neoliberalism Exceptionalism and the Culture of Uncare* (Bloomsbury Academic USA, 2021).
41 Aldo Leopold, *Almanach d'un comté des sables* (1949) (Flammarion, 2000), p. 178.
42 See Philippe Descola, *Par-delà nature et culture*, and *L'écologie des autres. L'anthropologie et la question de la nature* (Editions Quadrige, 2011).
43 B. Morizot, *Raviver les braises du vivant. Un front commun* (Actes Sud, 2020), and *Manières d'être vivants*. Préface d'Alain Damasio (Actes Sud, 2020).
44 Leopoldo Fulgencio, "Winnicott e o abandono dos conceitos fundamentais da metapsicologia freudiana", p. 225.
45 Aldo Leopold, *Almanach d'un comté des sables* (1949) (Flammarion, 2000).
46 See Hannah Arendt, *Nature du totalitarisme* (Éditions Payot & Rivages, 2018), pp. 129–39.
47 Hannah Arendt, *La condition humaine* (Editions L'Harmattan, 2001), p. 314.
48 See Chapter 7, "Unconscious Processes in Relation to the Environmental Crisis", pp. 81–92.
49 Günther Anders, *La menace nucléaire*. Traduction française de Christophe David (Le Serpent à Plumes, 1981).
50 Jean-Paul Engélibert, *Apocalypses sans royaume* (Classiques Garnier, 2013), p. 11.
51 See Amati Sas, "Mégamorts: unité de mesure ou métaphore?", *Bulletin de la Société Suisse de Psychanalyse*, 18 (1984), pp. 11–19.
52 See Hanna Segal, "Silence Is the Real Crime".

Part IV

Research

16 Development, Ambivalence, and Containment

Through the Himalayan Lens

Pushpa Misra and Jhelum Podder

The famous philosopher and environmentalist Arne Naess said that when people lose their habitat, the natural surroundings they have grown up in, they not only lose their place to live but also lose their identity. Questions regarding our relationship with nature are not new. They have been at the heart of classic texts by writers such as Naess, Leopold, and Carson; have given rise to disciplines such as ecopsychology, environmental ethics, and environmental humanities; and have also inspired new theories such as that of Biophilia. They have prompted empirical study. As psychoanalysts, we wanted to probe a bit deeper in exploring our relationship with nature, aware that this would require taking into account both conscious and unconscious factors. What is the place of nature in our internal world? How do the natural surroundings we grow up in become part of our identity? Indeed, what may happen psychically when connections with nature are lost?

This chapter describes our qualitative, interview-based, research project that explores these questions. Given that the human and the non-human environments become an intrinsic part of an individual from birth, we wanted to understand more about how the relationship with the natural world develops. Activities such as tree climbing, swimming in ponds, fruit picking, seeking shelter in forests and nurturing them, form bonds between humans and nature that are bodily and physical. Alongside human attachments, the natural world becomes closely linked with our emotions and often fills the role of substitute caregiver or significant other. This raises the question of whether non-human entities take on similarly powerful internal significance as humans in our lives.

We were curious to discover the quality of this significant relationship in human life through deductive qualitative research, expecting to find this relationship – an expectation backed by innumerable examples of people reporting their experiences of relating to nature.

We embarked on a small-scale study involving tracking the life stories of individuals, and analysing themes linked with their experiences with non-human nature. Taking both conscious and unconscious aspects into account, we sought to understand how people in our sample perceive

DOI: 10.4324/9781003634959-21

non-human nature and their often-ambivalent feelings towards it. We believe that answers to these questions can help us develop more effective strategies for people to understand their deeper relationship with nature, which is often obscured by other existential exigencies of life.

The most serious environmental problem we face today is the climate crisis which, as we know, is human caused. Understanding the values, emotions, beliefs, and attitudes we have towards nature, which on the one hand lead us to destroy it and on the other to protect it, would help us know what part of the human psyche to appeal to and encourage to grow and what part needs holding in check.

Sifting Through the Literature

While some very brilliant and insightful work has been done trying to understand the kind of relationship we harbour with non-human nature, space here prevents us from giving it its full due.[1,2] These texts have led to the development of ecopsychoanalysis, which combines perspectives from ecopsychology, psychoanalysis, other disciplines, and complexity theory, aiming to create a comprehensive, coherent, and encompassing psychoanalysis that integrates different ecologies. Whether ecopsychoanalysis will develop and whether a complexity theory approach will help solve the ecological problems, there is a truth in what Félix Guattari has said, namely that an integrated approach with other disciplines is a sine qua non for solving the ecological problem.

Research that comes closest methodologically to our research and also inspires the work we have done is Renee Lertzman's psychoanalytic fieldwork described in her groundbreaking *Environmental Melancholia* (2015). Lertzman argues that people's apparent apathy towards environmental degradation and loss is not due to indifference. Her deeply psychoanalytical study traces defences involved in this apparent apathy. Lertzman also traces ambivalence in the relationship with nature, as well as the human capacities for profound reparation.

Methodology

By "non-human nature", we refer to the natural environment surrounding us, excluding man-made elements such as bridges, highways, and mills, as well as man-made cultural artifacts like temples, churches, architecture, and sculptures.

Our research focuses on the role of nature in our psychic world and on the intricate links between the non-human environment and our psychological landscapes.[3] We have been aided in conceptualizing the complexity of these links by Anderson et al.'s "life-frame" of nature's values, which distinguishes between living from nature, with nature, in nature, and as nature.[4]

Our research aims thus necessitated inquiring into the meaning people ascribe to their interactions with, and experiences in, nature. Qualitative research was judged the most suitable method for this kind of inquiry. Because relationships invariably develop and change over time, we needed to delve deeply into the experiences of the participants vis-à-vis nature throughout the course of their lives, as part of their life stories. Indeed, a lifetime account was essential to capture the wholeness of experience in any given case. We decided to proceed with the narrative approach and the life story method as our instruments of enquiry.[5]

Narrative Approach and Life Story Method

Through narration, we understand not only the experience of the person but also its social context. Experience, therefore, is understood at the intersection of idiographic and sociological contexts. Within a narrative approach, people are embodied experience. They are not perceived as representative of their culture, race, or religion but as experiencing individuals with unique feelings, emotions, attitudes, and identities. The researcher becomes "autobiographically conscious of our own reactions". The researcher becomes wrapped up in the narration; he takes the story back with him, thinks about it, and writes it down. He also arranges the story in his own way and "restorifies" it.

To investigate the influence of nature on the human psyche, it was crucial to examine how individuals interacted with their surroundings throughout their lives. We required a comprehensive methodology that could capture the myriad ways in which nature impacted participants' internal development.

> A life story is the story a person chooses to tell about the life he or she has lived, narrated as completely and honestly as possible, what is remembered of it and what the teller wants others to know about it, usually as a result of guided interview by another.[6]

A life story is the story of the whole life, with its important events, crises, conflicts, emotions, and so on, usually guided by an interviewer – unlike oral history, which is the narration of the life of a community. Various social sciences use the life story method in this sense, and psychoanalysis especially uses it as part of its clinical methodology. As psychoanalysts, we found the life story method as the most appealing technique, partly because of its familiarity and partly because of its suitability for our inquiry. The common feature of all the stories is that they describe a situation where a crisis develops; there is a struggle by the protagonist and then the resolution of the crisis, where usually good wins over evil.

Robert Atkinson relates four major functions of all stories, including life stories. These are psychological (the narrator's evolving self and sense of identity), sociological (moral values, ethical standards prevalent in the society or community during that time period), mystical-religious (the storyteller may transcend the personal and "enter into the realm of sacred"[7]), and cosmological and philosophical. Regarding the last of these, our clinical experience has shown us that clients not only tell their stories but also come to understand the significance and meaning of important experiences and, by the time therapy concludes, seem to have arrived at a worldview, being able to bring some order into the scattered and disintegrated experiences of his life.

Thematic Analysis

We chose thematic analysis to analyse the gathered life stories. We observed certain themes in the stories, some quite similar or sometimes even identical to many individuals.

Thematic analysis (TA) is a method used to identify and analyse different patterns of meaning in a data set. The term TA is also often used synonymously with "content analysis"[8] to indicate methods which identify more interpretative forms of analysis based on recurrent themes.

Constellations extracted from the data set are usually found to involve affective, cognitive, and symbolic dimensions.[9]

Themes can be drawn from a theoretical idea (deductive) or from the raw data itself (inductive). We utilized a dual deductive – inductive and latent – manifest set of themes.[10] A good-quality thematic analysis relies on the subjectivity of the researcher, as well as their rigour and scholarship.

Database

To enhance our qualitative research (by providing valuable variety, introducing diversity, and enhancing its validity), we collected data from two distinctive, contrasting, and extremely diverse geographic regions of India: the mountains/hills and the plains.

Uttarakhand, a mountainous state in India, with its capital at Dehradun, was chosen as one of the hilly terrains for this study. Located in the northern slope of the Himalayan region, this state has observed massive development and environmental degradation in the last decade. We viewed Uttarakhand as an ideal location for our research.[11] The research was also conducted in the eastern foothills of the Himalayas, situated in the northern part of West Bengal.[12]

Kolkata and Delhi were chosen to represent the plains (the researchers are residents of Kolkata). These cities rank among the most polluted in India. We found it poignant to comprehend the residents' yearning for natural surroundings. This enabled us to engage in significant self-reflection.

Participants were selected by snowballing method. We interviewed individuals who were willing to share their experiences in nature. Most of our participants were enthusiastic about relating their stories to us. Inquiring about our research and objectives in conducting it, some believed that research alone would not be enough to save the environment and hoped that we would become activists for nature. Relocate.

Inquiry and Analysis

A semi-structured interview was employed to elicit the life histories of the participants. The interview served as a scaffold to help both the researchers and participants speak freely about their involvement with nature. The following were some of the questions asked:

1 How do you feel living in natural surroundings?
2 What is your story of growing up in this environment?
3 What role has nature played in your life and mind?
4 What was your childhood like?
5 Did you witness any change in the natural surroundings? What kind of changes?
6 What has your experience of these changes been?
7 How do you think these changes came about? Can anything be done about the harmful impacts?

The study participants were thoroughly informed about the aim and topic of discussion prior to consenting to participate. We commenced our venture only after they had given their written consent to participate in the study. Audio recording of the conversations was carried out with the participants' consent. Additionally, photographs were taken during the interviews. Through sharing meals and stories, we formed close relationships with the participants. We recorded our observations and reflections on each participant.

Conversations flowed organically as each unique individual story unfolded. Reading through the transcribed recordings, we could vividly imagine every adventure, concern, and historical event described. We decided to gather common themes which would cluster the most significant insights gleaned. However, space here prevents us from listing all the themes identified. And, some parameters of importance in the lives of people in hilly regions may not be present in the lives of city people.

Both urban and rural inhabitants were exceedingly cooperative when they learned of our work pertaining to ecology. The proprietors of the guest houses in which we resided in the northern and eastern foothills of the Himalayas went above and beyond to assist us by providing transportation, introducing us to their acquaintances, and persuading them to work with us. Amit Tomar, the proprietor of the Suwakholi guest house, expressed, "You all are endeavouring to preserve our ecology, while we are contributing to its degradation. Thus, it is imperative for us to assist you. Your team is performing commendable work." This also demonstrated the increased awareness among people in recent times about the significance of the environment.

Data and Analysis

Using coding and thematic analysis, we identified the following recurring themes related to nature and personal growth, which we have divided into two broad categories. The first category is "Development and Climate Crisis" which deals with tangible data. These can be objectively verified, if not with statistical data, then at least with the criteria of availability of essentials of life. "Relationship with Nature" is the second category, dealing with intangible and more subjective elements, which nature provides us without any demand or compulsion on our part.

1 Development and climate crisis:

   (i)   Construction of hotels, roads, buildings, dams, and so on;
   (ii)   Tourism;
   (iii)   Foreign job opportunities and their impact;
   (iv)   Flight from the villages;
   (v)   Unwillingness to work in agriculture;
   (vi)   Increased temperature;
   (vii)   Decreased snowfall;
   (viii) Construction of hydroelectric power projects;

2 Relationship with nature:

   (i)   Nature as playground;
   (ii)   Nature as nurturer;
   (iii)   Nature as provider of escape and refuge;
   (iv)   Nostalgia leading to melancholia in nature;
   (v)   Ambivalence towards development;
   (vi)   Craving for intimacy with nature.

Alongside this, our counter-transferences were also understood in connection with the research question.

Development and Climate Crisis

Despite considerable development in hilly areas over the past 30 years, life remains arduous. Most of our interviewees recounted a tough childhood marked by strenuous physical labour, destitution, and a lack of education and healthcare facilities. The statement below from DSP, 50, of Buranshkhanda, holding a good job, gives a typical picture of the conditions prevailing in most of Uttarakhand:

> First thing in the morning, we used to bring water from the jungle or from some spring. Then we would go to school (15 kilometres by

foot). After the school was over, we used to go to cut grass and bring it home for our cattle or to bring firewood from the jungle for fuel. Evenings came quite early here in our village. With the advent of evening, the whole village would plunge in darkness. We used to go to bed by 8/8.30 PM. We would get up early in the morning and follow the same routine. Amid all this work, we must go to school, even to our college.

In the hills of the eastern Himalayas as well, the conditions were not much different. GPS of Kurseong (aged 53) says:

Even at the age of four, we had to go to the jungle to fetch water from some spring. We were given small roundish earthen pots to bring water in them. I used to get up at 3 AM, and then would go to fetch water, then to collect firewood, light the oven – all these things we had to do. We didn't have time to go to school. I got admitted to the school at the age of eight in what we call "Dunga" class. A "dunga" class is where you are taught A, B, C, D of your vernacular. We had a wooden board, but no chalk. We used to learn the alphabets by writing with stones.

Agriculture was the main means of subsistence. People grew their own food; except for salt, nothing else needed to be bought from the market. Children often walked barefoot to school, wearing inadequate clothing even during winter. GPS says: "We did not have proper clothes to wear. Summertime was okay, but we didn't have clothes even in winter." RP (53) of Landour also said that they had only one set of clothes which they wore both at home and school. The school uniform was worn only on 15 August (Independence Day of India) and 2 October (the birthday of Mahatma Gandhi).

Pushpa detected a trace of bitterness in GPS's voice, as if he harboured some grievances against society and everyone in it. GPS was highly educated, having completed his Master's degree in Hindi literature – a challenging feat given the circumstances he grew up in – and is now well employed. According to him, times had changed. His 11-year-old daughter returns from school, tosses her school bag onto the couch, and heads straight to her room without doing any household work. Now there are roads, and with schools in close proximity, it is easy for children to walk to class. Transportation available in the area includes public buses, private cars, and both shared and private taxis. A frequent visitor to this area, Pushpa was pleasantly surprised to see the improved socio-economic status of the residents of Uttarakhand. In general, the homes we visited were equipped with almost all the fundamental necessities of life – clean and spacious living spaces, good clothing, and even refrigeration facilities. Additionally, the residents exuded a general sense of contentment and satisfaction.

Several key developments emerged from the interviews: the government of India prioritized the distribution of cooking gas cylinders to residents of the hills, leading to less destruction of forests; the construction of roads connecting major tourist destinations with main cities in the plains has resulted in increased tourism and a stronger local economy; hydroelectric dams have been constructed to provide electricity to the people of Uttarakhand; and more schools have been opened. With these measures taken by the government, the general standard of living of people in the hilly regions has improved. But healthcare is still significantly less developed, and higher education opportunities are limited to major cities in the plains.

Destruction of Nature and Climate Change: Hilly Region

Every participant had observed changes in the environment, noting shifts related to climate, their community, their general lifestyle, and value systems. RP said that during her childhood, Mussoorie used to experience plenty of rain, had adequate forest cover, and pollution was almost zero:

> Number of tourists was less, smoke from cars was nil and we used to have plenty of snow. Since the time tourism developed, trees have been cut down; there is little rain and more landslides. Snowfall also has become less. . . . Mussoorie was never so hot. . . . You must have heard that Gangotri glacier (the source of the river Ganges – the major river of India) is melting away very fast every year.

SJ, a mountaineer and adventure professional, also reported that the Gangotri Glaciers have receded due to pollution and tourism. The area, which was once desolate, lonely, and beautiful beyond description, has now seen a proliferation of cars, shops, ashrams, and other developments in recent years. RP continued:

> Mussoorie used to have lot of forest, wild animals. The wind was cold and pleasant. Now you feel the need of a fan. . . . Weather has changed a lot. Neither snow nor rain come on their scheduled times. This has happened because of cutting down the forest and people have done this because of their own selfish interest. . . . Because there are no trees, the soil has become loose, and more and more landslides are occurring.

RP further mentioned that, because of the increase in heat, new diseases are on the rise; because of increase in population, there is a large accumulation of garbage. Clean air is now a thing of the past, and tourists contribute to the city's uncleanliness by discarding leftover food, leaving behind residue, and urinating in public areas. This has led to an unpleasant odour throughout the region. Additionally, plastic wrappers and bottles have severely damaged the environment. Forty-five-year-old JR stated that previously, the area used to have about five feet of snow in winter. These days, it is hardly half a foot and that also clears within hours. Earlier, due to heavy snowfall, transport used to remain suspended for six months. Now, the bus route remains active. Business goes on. A table would be set up anywhere on the roadside, and tea, coffee, fritters, and so on would be sold. He asserted that this has happened due to tourism. Throughout the year, tourists are coming from all over the country.

With access to technology, people's lifestyles have changed. There is no longer any need to accumulate firewood for the rainy season because people now have gas cylinders. The ladies of the house no longer have to do the hard work their parents once did. There are now gadgets for almost every household work. Twenty years ago, this land was not deemed worthy of purchase. However, all the available land is now sold out. Individuals from the plains have migrated here, resulting in a population increase. Additionally, hotels have been erected, providing potential sources of income for residents. Thus, the general populace is in favour of the development.

The state of the Eastern Himalayan region is similar. Increased forest coverage has led to more rainfall. Recently, Darjeeling has finally had a snowfall after many years of a dry season.[13] However, the widening of forested areas has impacted on agriculture in the region. Land that once yielded crops such as corn, millet, rice, and pulses is now forested, housing deer and monkeys – animals previously unseen. The climate has undergone noticeable changes, with rainfall occurring when least desired and droughts during crucial periods.

Urbanization in the rural areas has altered traditional lifestyles and community dynamics. Homestays, now ubiquitous, have led to erosion of community bonds and mutual empathy.[14] Mobile technologies are pervasive, with people constantly occupied with their devices, even during gatherings. With economic prosperity, traditional clay houses – cool in summer and warm in winter – are being replaced ubiquitously by cement dwellings. The younger generation has minimal tasks or responsibilities, spending their free time attending school, playing games on their mobile devices, or watching television. This has led to diminished connections with neighbours and friends, causing a potential impediment to their holistic development. Greater forest cover has resulted in a decrease in landslides. Nevertheless, according to GPS data, this perceived development has disrupted the ecological equilibrium.

The government's new industries and hydroelectric plants have loosened the base of the mountains and have made the ecology of the region very fragile. The hills from one part of Sikkim to Siliguri where there are hydroelectric generation plants are prone to very frequent landslides. The impact of hydroelectric projects was emphasized by DPS of Uttarakhand. He informed us that his family was in agriculture but now they have reduced it to a minimum. The reason is a disproportionate increase in the number of animals, such as deer, monkeys, and pigs, that destroy the harvest.

In general, people don't want to do agricultural work any more. There is large-scale flight from villages to provide for better education and health care. "90% of the people leave the village because of lack of health facilities", DPS iterated. According to him, forest area has increased because of the availability of cooking gas. Better roads connect villages, and more dams are being constructed in Uttarakhand. Numerous hydropower projects have been initiated by the government. DPS said,

> I understand the need of the country because these dams do not provide electricity only to Uttarakhand but to other parts of the country as well. But this does not mean that all dams are to be constructed in one place. . . . Even before the construction of one dam is over, another one starts. . . . They have stopped the flow of the river for 300 meters. The entire environment is destroyed.

To construct dams and widen roads, hills are being blasted, which loosens their roots leading to frequent landslides which uproot trees and plants, causing further environmental degradation. In this season alone, he informed us, there have been at least ten incidents of landslides in the Chamb area. Hills are also being blasted to construct hotels, which are proliferating in the area.

Pushpa: You talked about tourism and the construction of hotels. That brings economic benefit to you. If tourists come, hotels must be built and hills must be blasted. Yet you also feel troubled about the destruction of the environment also.

DPS: We need tourists, but they should behave responsibly. They should not cause any damage to our environment. Trees and plants should not be harmed. Plastic bottles and plastic wrappers are strewn everywhere. Plastic is dangerous for health and environment. It eats up the soil . . . they should think about these things, and we should also tell them about these things.

DPS was evidently agitated. He also seemed quite knowledgeable and well informed about the environment, pollution, and its detrimental effects on health. Tourism was both desired and resisted simultaneously. Electricity was wanted but not at the cost of environmental destruction. Everyone we spoke to was aware of the pollution, the blasting of hills, the increase in hotel construction, the increase in traffic and carbon emissions and everything that leads to climate change. But they also wanted the benefits of these changes in development. The very frequent blasts to make room for development were thought to be responsible for this devastation.

DPS's anxiety about landslides due to blasting of the hills is justified. Experts are of the opinion that:

> Mushrooming of hydropower projects existed in Uttarakhand. There-fore, it is called "Urja (power) State". Every river and its tributaries have more than one hydropower project. About 32 new hydropower projects are being constructed. Recently, the Supreme Court of India has banned the construction of 24 already planned hydro projects in Uttarakhand.[15]

Most rivers in India are fed by Himalayan glaciers:

> Even, a small earthquake may be a potential danger for these glaciers. High climate variability and change were noticed in the Uttarakhand Himalaya during the past decades. An increase in temperature was noticed by the author.[16]

According to a report by the Wadia Institute of Himalayan Geology, due to the warming and melting of glacier snow, the number of glacier-fed lakes has increased from 39 in 1976 to 217 in 2011.[17]

The important question, however, is whether the state of Uttarakhand will be able to bear the consequences of these interferences with its ecosys-tem. Expert opinion is that

> human interferences in terms of mushrooming of infrastructural facil-ities, mainly construction of hydropower projects, along the highly fragile and vulnerable landscape/river valleys are catastrophic, which need immediate measures. Otherwise, the future consequences will be horrible.[18]

The experts think that it is important to give a serious thought to our devel-opmental policy, so that the fragile Himalayan ecosystem can be saved without hurting the welfare of the people of Uttarakhand.

SJ, a trekking guide, stated that agriculture demands a considerable amount of physical labour. Consequently, fewer people are inclined to enter the agri-cultural sector. On the other hand, hotel jobs offer better pay and require less physical strain. SJ further mentioned that he plans on venturing into the hos-pitality industry once he can no longer go on treks. He might even consider establishing an agency. However, he did not contemplate pursuing a career in agriculture. He confirmed a rise in landslides and a constant underlying sense of destruction. However, informed by his experience in adventure, he expressed confidence in his ability to rescue numerous individuals.

TS of Kalimpong in the Eastern Himalayas asserted the reality of climate change. Peaches typically bloom in April, yet they had already done so in

October. Similarly, oranges are currently in bloom, even though it is not the usual season. He showed us these fruits in his garden, which he had diligently wrapped in plastic. There was a solitary occurrence of unseasonable fruit blooming in Kalimpong that was reported by BKS.

So-called development has taken place in urban areas as well. There is mindless construction of high-rise buildings, flyovers, roads, shopping malls, restaurants, and so on. Air pollution in metropolitan cities is very high. In some cases, conditions in urban areas are worse than in the hilly regions.

Overall, we were pleased to see the economic progress of the hill communities. While some may question the extent of the development in these areas, people have sacrificed intangible benefits for tangible ones. Only time will determine whether this trade-off proves worthwhile or not. Pushpa, as a frequent visitor, was particularly struck by the changes. Most people we spoke with expressed sorrow over the increase in temperature, reduced snowfall, and declining agricultural activity, all of which threaten their traditional way of life. It was apparent that the hill residents had mixed feelings about what they referred to as "development". While they acknowledged its necessity, many felt uncertain about it.

Ambivalence was evident within the narrative data, indicated by fluctuations in opinions from positive to negative, particularly towards the industrial sector. Ambivalence also encompasses a significant aspect of ecological participation that perceives industry as challenging, encompassing aspects of health and harm, specifically those that cause worry but are not remedied.[19]

Lertzman argues that ambivalence towards industry and the environment leads to a melancholic response to ecological degradation. Protesting or expressing anger directed at the source of this loss may not be feasible due to concerns about possible consequences such as unemployment, social isolation, or being seen as unappreciative of the benefits of industry. Ambivalence and melancholia therefore constitute a set of defensive strategies that shape action, reaction, and involvement.

We felt a sense of disappointment, as if we had expected something different from the locals. It wasn't solely the devastation of the environment that brought sadness, but something more. Perhaps we unintentionally projected our own expectations onto them, maybe believing that those living in hilly regions should be content with less. Pushpa's disappointment was more evident. Perhaps she was expecting the attitude and values in the hill people that she had witnessed and encountered 30 years ago – that they would avoid greed and be satisfied with less. It proved challenging for us, as middle-class urbanites, to grasp the significance of having access to electricity for children's studies, running water within the home, the assurance of two square meals a day, and a clean, tidy dwelling – privileges that we assumed as the norm. Regarding greed, does this not also hold true for individuals residing on the plains?

Indeed, it does to a far greater extent.

Relationship with Nature

Our earlier narrative gives a glimpse of the role nature played in the lives of the hill people; nature was part of their daily lives – indeed their survival depended on it. In contrast, urban children generally have only occasional encounters with nature, mostly through pets, gardens, and trips to areas known for their natural beauty. Poverty exists in cities, too, especially in developing countries like India. Urban poverty is much uglier than poverty in hilly areas. Homeless people, beggars, children being trafficked, and so on. These are common stories of all big cities. In contrast poverty in the hills is not as harsh in these ways. We have data from poor and middle-class people who are mostly educated, and some highly so.

AB, a 32-year-old ceramicist, stated:

> I used to enjoy storms . . . when the storm comes while we were play-ing on the playground, we used to run for home. Generally, the storm during the month of April used to come in the evening. To me this experience was of joy, not of fear or panic.

There were small pleasures – like hearing rain falling on rooftops, sit-ting in the park experiencing the peace, seeing the flowers blooming in the garden, catching butterflies, looking at the mountains from the windows of a homestay, or simply gazing at a big tree – that brought immense pleasure and happiness. Those living in neighbouring villages would help their parents with agriculture during rainy season. SP mentioned:

> My most favourite season is spring. The wind is very sweet, the cuckoo calls incessantly, the wind gives a special good feeling of wellbeing, it's neither too hot, nor too cold. It has a very special flavour – can't express it in words . . . There was a "Kadamb" (Neolamarckiacad-amba) tree in our courtyard. It had a very nice, sweet smell. With western wind the smell used to enter our rooms. . . . This was before my marriage. Then the tree was sold for my marriage.

Pets seem to be a very important part of urban children's connection with nature. IM said that they used to take care of six or seven stray dogs: "I think, it definitely helps me to be a better person. Just don't be cruel – the basic bottom line is what kind of world you want to leave for your children."

Some of our urban participants said that they are extremely fond of trees and plants. CBG said that in his office he has many plants in pots and that he takes very good care of the them, cleaning every leaf with his own hands.

Nature as a Playground and Bonding With Nature

The most exquisite manifestations of the connection with nature were arguably those in which the participants recounted childhood moments spent frolicking in its embraces. Oftentimes, while recounting their mischievous antics within natural environs, they would chuckle and radiate pure joy. The vividness with which they painted pictures of their playful bond with their surroundings was a testament to how deeply cherished these memories truly were. Their cheerful vocal inflections and lively facial expressions consistently revealed an unmistakable relaxation into a more carefree state of mind.

JR (45) was a mischievous child who caused his parents distress by climbing from tree to tree in the forest and playing with the soil in the fields:

> We used to play with the soil the whole day. Our whole body would be covered in mud. Our parents rebuked us, but what can be done? Children will be children!

The enjoyment of unrestricted outdoor freedom, where children can run and play without concern for repercussions, unencumbered by homework or duties, is a delight for youngsters everywhere. This reinforces the notion that the developmental stages of children around the world are remarkably similar. The human mind, in a regressive state (meaning a reconnection with childhood experience), often fantasizes about relatable childhood events and anecdotes, especially those related to play. Nature, a pervasive and inclusive force in everyone's life, fully nurtures the imaginative potential of young, prodigious minds. This is only possible due to the versatile and colourful traits of the natural world, which children can easily relate to. The sense of sanctuary, containment, and envelopment all relate to a physical and visceral feeling of being enclosed within something that provides protection.[20] Therefore, this experience of safety helps explain why it is so much easier to adopt nature as a playground.

A fascination with mud and mess, which was also expressed by JR, was apparent in another participant's life story. SP, a 39-year-old homemaker, shared childhood anecdotes of playing in the rain. This suggested the universality of the anal stage, as described in Freudian psychosexual development.

> There are no dirt roads anymore in the village where my childhood was spent. All of them have become cemented. In our childhood, it used to get muddy during the rainy season. While going to school we used to get drenched and slip on the mud.

Jhelum: Did you do it purposefully?
SP: Yes! It was a game for us . . . it was a different kind of fun going to school back then (giggling)!

Further into her recollection of the past, she continued in the interview:

SP: My grandparents and my mother built a room completely by themselves with clay. Even me and my sister used to jump in to participate. Those days are just lost.

In recalling memories of nature and childhood, the affective qualities were overwhelmingly associated with a self-state of innocence or a perception of the world as a safer place than it is now. Feelings of innocence are related to feelings of safety and containment.[21] This also helps explain why so many of the participants take shelter in natural environments during difficult and painful life situations.

Children in the hilly area must also perform a lot of household chores. They walked to school through the forest. While this was hard work, it also had a playful aspect. Nature served as a playground for them. RP says:

As a child, I did not have a single toy, nothing to play with, nor did we have a TV. . . . It was the jungle that gave me a place to play. We used to play in the jungle, we used to swing holding the branches of the big trees. I think that the good health we enjoy today is because of our play in the jungle.

RP later took a course in rock climbing, camping, rafting, and, of course, trekking. She was an adept rafter and rock climber and had accompanied many tourist groups to various places.

BS of Rautu Ki Valley was eloquent about his childhood experiences in nature. He said that forests were very close to their home because the population was sparse. They could see a variety of colourful, beautiful birds. He says:

We had cows and goats, and my parents used to go to the forest to graze them and we used to accompany them. We would skip school to go to forest with them. If they didn't take us with them, we would cry and force them to let us come with them. . . . We would bring fodder for our cows.

BS said that for them, there wasn't any difference between the jungle and home. Their home was amid the jungle, as he explained, "We lived in jungle and jungle is our home." The jungle was their playground. They used to make swings on trees, and swinging was one of their most enjoyable games. During summertime, schools closed early, around 1 p.m. All the children would go to the forest to cut grass for the cows. They used to sit together, gossip together, play on the trees, and keep themselves busy until it got dark. In this way, the forest became their playground.

Several recent studies have highlighted important features related to our interaction with nature.

> Children deprived of playing in the green spaces lose a lot more than fun. The research suggests that they will not reach their potential for motor skills, balance, creativity, attention, social and emotional skills, and spiritual fulfilment. These crucial life skills and proclivities need to be developed while the brain is still highly malleable.[22]

Could it be that the foundation of meeting challenges and bearing an enormous amount of hardship is laid in the childhood play of hilly people?

Longing and Nostalgia for Natural Surroundings

It is widely recognized that childhood represents one of the most treasured phases of human life, barring instances of trauma or suffering. This period evokes a sense of magical omnipotence and exuberant existence. The boundless freedom offered by open lands, the towering grandeur of mountains, and the spirited flow of streams all serve to invigorate the young and curious psyche. Memories and associations of nature and childhood frequently seem intertwined and inseparable.[23] Such sentiments were also echoed by our participants as they recounted how nature shaped their formative years.

WL, a 35-year-old doctoral candidate in Buddhist Philosophy residing in a village in Kalimpong district, fondly recalls venturing into dense jungles to collect an assortment of fruits and banana trees, while also gathering grass to feed their bovine friends. He expressed:

> The surroundings were so beautiful, we used to go to the jungle with our friends, you know, howling sometimes. And we used to gather the jungle fruits.
>   . . . I used to bring banana trees in my home, because the banana trunks were given to the cows as food.

WL was able to engage in both introspective reflection and imaginative pursuits. He voiced:

WL: I used to write little poems and short stories, as well. For the people like me hilly region is very good.
Jhelum: What do you mean by "people like you"?
WL: I mean . . . you see clouds, you see blue sky, you see mountains, rivers, everything. And you can just write it down, alone, in the forest or on the top of hills. You can concentrate. Concentration is so high when I am in the forest.

Immersing oneself in a lush and bountiful environment can prove to be an immensely fulfilling journey for those seeking to unlock their innermost potential.

Residing in urban areas presents a distinctively dissimilar experience. Over the course of 23 years, India has seen an increase of 67.7 percent in annual particulate pollution rates. To add to this alarming statistic, from 2013 to 2021, India was responsible for a staggering 59.1 percent surge in global pollution levels. The sentiment of being deprived of fresh air among verdant landscapes is frequently voiced as a painful struggle endured on a regular basis. The feeling of loss is emotionally linked to the loss of something undefined and preceding adult understanding: a time of carefree purity.[24] As a result, it has become customary for city dwellers to seek solace in remote locations within the country, away from the chaotic hustle and bustle that characterizes their dry, work-life routine.

In the following excerpt, GG, a 55-year-old social worker, discusses her childhood residential area's lack of greenery and how her relationship with nature evolved over time. Initially deprived of interaction with the natural world, she now experiences an intense desire to be among mountains as she grows older.

GG: I recall having very small plants in our house where I grew up. But they were not of any importance to me and the surroundings completely lacked greenery. . . . But as I aged, I constantly longed to join organizations that work in and for the environment.

She further said:

GG: I never liked mountains before but now for some incomprehensible reason I have this unexplainable pull towards the hills. The vastness of the mountains and the peace around it completely merges within me.

Nature Providing Calm

Different people attributed different qualities to nature at different times such as calming, serene, restorative, as well as fearful. It is possible that nature does not appear as something special to people living in the hills. As BS said, "There is not much difference in the jungle and our home. The jungle is our home, and our home is in the jungle." But urban people must make a special effort to meet with nature, to stay there, and to experience the feelings it aroused by it. IP, for example, says:

A river is very calming. . . . We were walking together and there was a disharmonious silence because we were arguing about something. As we approached the bank of the river, we both fell silent. I still remember, it was a night of full moon, dusk was gathering, a mystic colour seemed to cover the atmosphere. I remembered a line from a poem, "It's a moon that dances in the courtyard of the dead".

GG was deeply involved with nature. There were elements of spirituality – not religion – in her love for it. She said, "There is that One sitting in the sky who runs this universe, somewhere I feel as if I can see Him in nature. When the rain drops fall on my body, I feel these are His blessings and internally I become peaceful." Raindrops, flowers, and butterflies all affect GG very strongly in a very positive way. If she is feeling depressed and the sky is also gloomy, depression intensifies. But suddenly, sunlight enters the room, and the depression vanishes. "It seems as though someone listened to my thoughts." The openness, the expanse that nature provides is also very liberating for many. For AL, space is very important.

> Whenever I see a patch of greenery, I feel thankful. . . . I feel pain when trees are cut down. During childhood we used to go into the forest. Now that forest has been cut down to make way for a flyover. . . . Things have changed so much.

SM is a mountaineer. During her training, she was given a task of spending a night in the jungle with some coffee, noodles, and a plastic cover. After she overcame the initial fear, a chain of memories started crowding her mind – some happy, some sad, and some that made her angry. It was as if her mind were re-evaluating all these memories. Eventually, a realization dawned on her: there is no point in remaining angry and severing relations with anyone. Her mind became very soft and very expansive. She thought that after returning, she would forgive everyone and repair her broken relationships. Her heart became tension-free and light.

> *Pushpa:*   It seems that the atmosphere of the forest, the one-million stars you could see through the small gap in your plastic cover, somehow brought an expansion in your mind. . . . Your mind became much broader. You could forgive the small shortcomings of human beings.
>
> *SM:*   Not only the vices of others. I could forgive myself also. You can say I felt more confident about myself; I understood many things about myself and felt that I have suddenly grown up.

SM also said that she has a very deep relationship with trees, rivers, and mountains:

> When I saw the Neelkantha mountain at Badrinath[25] for the first time . . . a very strange feeling overcame me. . . . I had a feeling that

whenever and wherever I die, I would like to see Neelkantha just before I die.

As a mountaineer, SM has also seen the dangerous face of nature – heavy snowfall, avalanches, melting glaciers where every step could bring about death. Yet the mountain, she says, has helped her to know herself.

Whole day we have climbed, we did the summit and then started climbing down. . . . I have seen the approaching dawn standing at a height of 20,000 feet, the play of light. . . . Picture, camera? There is no camera in the world that can capture the beauty of that scene.

Near the source called Gomukh (meaning "mouth of a cow" because of its shape), the water of the Ganges (the largest river of India) is crystal clear. One can see all the stones on the riverbed. She said:

I experience a very strange feeling when I see that crystal clear water. I feel a strong desire to go down under the water and lie on those stones. For some reason, I feel that it is so peaceful there. This feeling I experience when I am, for some reason, particularly upset and feeling deeply sad.

SM also said that she has learned quite a few things from the mountains; she has become quieter and more accepting, as well as more fearless and confident. The mountains have brought out certain hidden potentials in her. We have presented SM's account only briefly. The place of nature in her mind is very deep and intense. It has transformed a quiet, isolated, brooding girl into a courageous, brave, and team-spirited person.

Nature as a Refuge

When discussing the challenges of a bustling city life weighed down by numerous responsibilities, individuals often seek refuge in a trip to the seaside or countryside. Removing oneself from daily concerns is a convenient, albeit temporary, solution. The present research notably highlights how we unconsciously turn to nature as a healer for many of life's worries and sufferings, and it is not confined to merely avoiding a fast-paced urban lifestyle but rather is embraced by those who lead a serene and unhurried life, waking up each day to the breathtaking beauty of rolling hills and verdant forests. GG talks about finding solace through nature in her loneliness:

Pushpa: This personal transformation where you feel you are a part of nature yourself – how did that happen?

GG: There was a time when I was very busy. I had a kid after marriage and then pressure increased in my job. I also had to take care of my ailing parents. And then suddenly there was emptiness. My parents died, my daughter grew up . . . and in such phases of life, one needs some shelter . . . nature entered this empty space within me. I feel very content being in our small garden. I am happiest when I observe a blooming flower.

For DD, nature acted as a healing force during her childhood. She encountered health issues stemming from a congenital heart defect, resulting in recurrent illnesses:

DD: In my childhood, I could not sense the impact of nature, as I was preoccupied with health issues. I had to go through bypass surgery as there was a hole in my heart.

Jhelum: Oh!

DD: Once I survived, I began to feel that nature kind of helped me in healing. There was no rush, no pollution. It had a very calm environment. There were many trees providing ample shade and cool. . . . I have a strong attachment to my home because of the environment that surrounds it.

At the conclusion of the interview, DD revisits the topic of nature, underscoring its significance in her life:

DD: Nature actually helps when I am stressed. Life was not easy when I arrived in this campus for my master's. I still struggle sometimes, especially with respect to being a resident here. . . . To reduce the pressure, I used to just go out for a walk in the tea garden. This was extremely healing for me. . . . Nature has been very instrumental in healing a big part of me.

Isis Brook cites four studies performed on various groups, including those accused of domestic violence and those suffering from long-term mental health issues. The studies aimed to find out the effects of exposure to nature on the participants. The tasks were simple: having a picnic in a garden, planting trees, or playing with pets. The results were remarkable.

For example, someone with long term mental health problems who has come to see themselves as unable to do anything positive and as a burden to everyone was emotionally moved and found a spark of something positive to work from. When she discovered that the seeds she had planted in the seed bed she had prepared had germinated, the revelation that she could bring about something good in the world led to an increase in self-belief and confidence.[26]

Most of us, including Freud, have talked about our relationship with nature in terms of victory and subjugation, master and servant. We do not completely deny this, but just as nature was fearful since time immemorial, it has also been seen be loving, caring, and nourishing from time immemorial. When we asked our hill-dwelling participants whether they felt angry with nature for bringing so much trouble into their lives through floods, landslides, and so on, most of them responded that it was largely the fault of humans for overburdening nature. So the devastations had to happen. It requires a deep understanding of the ways of nature to be able to give such a response.

While DD was being healed from physical ailments and academic stress, RP often sought solace in the forests during times of emotional turmoil:

RP: When I had fights with my siblings or my mother and felt angry with them. I would run to the forest as far as possible, irrespective of rain or sun. And once I reached deep into the jungle, I would scream loudly. This would give me peace immediately. I would return only after that. . . . When I used to stay in Ireland, whenever I missed my family, I would go to the Atlantic Ocean which was 3 ½ miles away from where I used to live. This was one of my favourite places.

The discussions with our participants revealed that nature reflexively takes on the role of a nurturer for humans in need of healing. Nature's containment, whether for physical illnesses or traumatic experiences, instils a sense of calm and peace in those who are distressed. Innocence and freedom can also be achieved through the perceived existence of boundaries and containment.[27] The vast expanse of natural surroundings possesses an almost divine ability to alleviate human pain and sorrow. This could suggest why the majority of participants demonstrated a profound affection for their hometowns, nestled in the bosom of nature or surrounded by its various elements.

The natural environment and outdoor experiences evoke feelings of innocence, purity, containment, and sanctuary. Sanctuary refers to a sense of safety and comfort, signifying a positive environment.[28] Winnicott's concept of being held is related to this idea, as is Bick's notion of ontological containment, which envelopes an individual like skin.[29] The tranquillity and refuge offered by nature were documented in numerous participants' accounts.

People perceive their connection to the planet through their bond with their mother. While it may seem anecdotal, the notion of the mother-child relationship is profoundly important in shaping an individual's worldview. The first encounter with the mother or caregiver serves as the basis for one's relationship with all other objects, including the earth.[30] It is conceptually plausible that humans would extend their initial relationship onto their perception of the world. This helps explain the use of the phrase "Mother Earth" across various cultures.[31]

Bonding With Nature

There is no doubt that people living in the hills are more closely connected with nature, and their bond with it is also very strong. RP stated that if she were asked to choose between living among the grand buildings of a city and the environment she currently lives in, she would most certainly choose her present environment – an environment with rivers, mountains, and so on. She said:

> I think here we have life. In big buildings I feel suffocated, here I can breathe . . . I love tree and plants, I can live freely, hear the songs of birds, I can touch trees. This is my real life.

SJ mentioned that he grew up amid mountains, rivers, springs, and forests, and that he does not feel comfortable living in cities. When asked, "How do you feel when you look at the mountains?" he said:

> (While trekking in the hills) It's very peaceful here, very joyous! . . . There isn't the tiniest of sounds. The sleep there is so blissful. . . . Sleep amid the mountains is deeply peaceful. A two-hour sleep in the mountains is comparable to 6 or 7 hours of sleep in the villages. Even after two hours' sleep you feel refreshed.

JR said that he still looks after the bamboo forest his parents developed and nurtured. He has never cut down a tree in his entire life and has similarly instructed his wife and children never to cut down a tree.

Many of our interviewees said that they would not like to live in big cities. Even if they have to move to cities to earn a living, they would all prefer to return to their villages after retirement. They also regularly visit their villages on festive occasions. DS said that they don't have to think about their surroundings in any special way. The whole environment they live in remains with them in their hearts. WL said:

> Whenever I would go far from my village, even in the . . . town, I remember my village and the surrounding of that village. Because, from my village, you know, we can see the full range of mountain Kanchenjunga clearly and that, I think, has some sort of spirit, some sort of positive vibes that gives some energy.

There is no doubt that, just as the child internalizes the significant people in their life through constant and close interactions, they also internalizes their natural surroundings – the hills, rivers, springs, and so on. This alone explains why almost all our participants said that they don't have to think of nature separately, consciously, or deliberately. It is there with them all the time, as the background to everything they do. An unconscious identification with natural surroundings occurs spontaneously which often becomes part of our psychic landscape.[32]

It is not always possible to discern the feelings one experiences being in the natural surroundings.

Jhelum: What does that place give you? You go back . . . how do you feel?
WL: I cannot express the feeling 100% in words, but . . . it is like, I think seeing the mountains and blue sky, the hills. The more these come to your vision, the more you feel relaxed. . . . You see clouds, you see blue sky, you see mountains, rivers, everything. And you can just write it down, alone, in the forest or on the top of hills. You can concentrate. Concentration is so high when I am in the forest.

Different aspects of non-human nature affect people in different ways. Flowers blooming seasonally, the mountains, rivers, springs, the blue sky and the moonlit night, the quietness, the solitude – all affect the people living both in the hills and in cities.

Often, a spiritual connection with nature develops to the extent that it becomes part of one's identity. We believe that most people from the hills expressed this in their interviews when they stated that the environment is always with them; they don't have to think of it separately. In the case of urban residents, this happened sometimes. AP uttered, "Initially, I used to think that the soil is something different from me. But now I think it is part of me. I mean, I have come to that comfort zone where I can think it is part of me. Wherever I go, it will be with me".

People often have transformative experiences when they are deeply connected with nature. SM, our mountaineer participant, had quite a few transforming experiences which have deeply affected her personality. TS of Kalimpong, a highly educated man in his mid-40s, comes from a wealthy family. His father had a lot of land and had planted many trees – mostly bamboo trees which grew into a forest. Many people advised him to sell the trees, saying it would bring in a substantial amount of money. But TS refused. He has preserved all the trees and showed us the bamboo forest he has helped create. Their land is slightly downhill, and TS often goes there to spend a few hours alone. He said, "My wife asks me what I do there alone." I tell her, "Come, spend a few hours here and see for yourself." He relaxes there, staying alone amid nature. Sometimes, he even spends the night there. He said, "When I come back, my face changes. There is a shine on my face, and that is why my wife asks this question." Forests and mountains can be sources of deep spiritual experience.

People in urban areas also develop deep connection with nature, even though they do not encounter it very often. Apart from mountains, they also become deeply connected with animals, trees, plants, rivers and so on. GG said:

> I am thinking of sponsoring a baby elephant when I retire. My colleagues laugh at me . . .

Pushpa: Of all the animals, why did you choose a baby elephant?
GG: I had read that elephants are intelligent animals. Their huge body, their loyalty to their group – for some reason I have a craving for this. . . . When I see a baby elephant, I just want to hug it, caress it Intelligence – yes, it attracts me very much.

She further said:

> It seems to me that people seek some kind of shelter. Some seek it in books, some find solace in other people's company, I have probably taken shelter in nature. . . . I feel profound peace in nature. I feel that so long as your parents are alive, they accept you as you are. After that where would you find this kind of shelter? Everyone else judges you, nature does not.

Kohut has said that we need a self-object throughout our lives. Does nature sometimes become our self-object, taking the place of our original self-object from childhood?

AL said that she had to leave her place of origin which was in North Bengal in a semi-hilly area and now lives in Kolkata. However, she would like to remain connected to the place where she spent her childhood. She said:

> Yes, I mean that environment – the trees, the animals, the birds, they are very important to me – not because our existence, our survival depends on our ecosystem, not in that functional sense, but from an emotional point of view. I feel if they are not there I also will not be there emotionally.

Sometimes a very deep and emotional relationship with non-human nature develops because of certain special circumstances. SR describes her unique connection with nature. During the Bangladesh Liberation war, they took shelter in a forest. They remained there for two months. In her own words:

> We would hide on the top of some tree having many branches and thick foliage or behind thick bushes . . . there were big snakes hanging from the trees. But they never harmed us even if we touched them accidentally.

Pushpa: So the jungle was like your mother. It nourished you, it protected you, contained you, held you.

SR: Yes, this forest was exactly like our mother. If it were not there, none of would have survived.

SR is very fond of trees. She now has a house of her own, where she has grown all sorts of large trees and takes very good care of them.

AL provides us with a strange but very deep psychological connection to the house she grew up in. She misses the house itself – its physical arrangement – the very space. It is almost a compulsion for her to visit her native home during the festive season.

AL: I mean, the house had become almost like a mother to me. I think of the house mostly, not of the people whom I have left behind. Even now when I go there, my priority is how much time I can spend in the house . . . it is possible that my mother remained too busy and the house gave me protection.

Pushpa: So you felt very secure in the house.

AL: Yes, that I did.

AL has resented leaving her house in North Bengal. She had to move to Kolkata for her education, but it was a sudden decision made by her parents. She has always felt that she was uprooted from her home. She also informs us that she feels a strong sympathy for refugees – those who have lost their homes – such as the Bangladesh refugees and those from erstwhile East Bengal.

Our psychic space is populated not only by the representations of significant people in our life, but non-human nature also occupies a large part of this space. People experience pain when the "kadamb" tree in their courtyard is cut down or sold to someone else, when the river flowing beside their village is blocked to construct a dam, or when a hill is blasted away. It is not merely the physical consequences of these activities that cause anguish; it is experienced as a personal loss, like the death of someone close and dear.

Conclusion

Our study is qualitative in nature and based on a limited sample. Hence, we do not aim to generalize our findings. However, we do gain some insight into the questions we began with, namely understanding more about the role of the non-human environment in our psyche. We received very rich responses from our participants which, in some cases, gives us new insights about nature-human relationship and, in others, confirms our nascent ideas about this relationship.

Borrowing from Anderson et al.'s conceptual framework, that distinguishes living from nature, living with nature, living in nature, and living as nature, we will focus only on the first and the last – though all the four seem to be interrelated in some ways. Living from nature emphasizes the instrumental value of nature – the use we make of it to satisfy our needs; while living as nature emphasizes that "nature matters because it constitutes people physically, mentally and spiritually through our relations, our kinship, our interdependence".[33]

Our data expresses that current policy leans more towards living from nature. The entire climate crisis, including the change in weather experienced by the people of Uttarakhand, stems from the utilisation of nature only as a means to satisfy human needs. We are not unaware of this tendency in ourselves. It is important to clearly distinguish between our needs and our greed. We are reminded of Mahatma Gandhi's words: "Nature has enough to satisfy our need, but not our greed." It is probably important for us to develop a philosophy of ecology – an ecosophy – that will guide our policies and actions related to nature so that there is a balance among all four aspects of our relationship with it.

The final frame, living as nature, was our main concern, and our data shows how deeply our psyche is connected with non-human nature. It really constitutes us by providing shelter and refuge, by teaching us to overcome adversities, by becoming a substitute parent (as seen in the cases of GG, AL, SR, and SM), and by giving us solace when we are troubled.

In conclusion, we think of Harold Searles' book in which he, so long ago, emphasized the extremely intimate relationship between nature and human beings. It is important to recognize that this is a relationship with its own independent value, contributing immensely to our personal growth. Probably we also need to realize that when you look out at the immense expanse before you from a height of 20,000 feet, you do not necessarily regress but instead emerge from your infantile narcissism and recognize your proper place in the vast universe.

Notes

1 Arne Naess, quoted by Richard Langlais, "Living in the World: Mountain Humility, Great Humility", in *Deep Ecology for the Twenty-First Century*, ed. by George Sessions (Shambahala, 1995), pp. 198–99.

2 See in particular Harold Searles, *The Non-Human Environment in Normal Development and Schizophrenia* (International Universities Press, 1960); Joseph Dodds, *Psychoanalysis and Ecology at the Edge of Chaos: Complexity Theory, Deleuze, Guattari and Psychoanalysis* (Routledge, 2011); Sally Weintrobe, *Engaging with Climate Change: Psychoanalytic and Interdisciplinary Perspectives* (Routledge, 2012); and Renee Lertzman, *Environmental Melancholia: Psychoanalytic Dimensions of Engagement* (Routledge, 2015).
3 See Weintrobe (2012) for the concept of psychological landscape and nature as an internal object.
4 See C. B. Anderson et al., "Conceptualizing the Diverse Values of Nature and Their Contributions to People", in *Methodological Assessment Report on the Diverse Values and Valuation of Nature of the Intergovernmental Science-Policy Platform on Biodiversity and Ecosystem Services*, ed. by P. Balvanera et al. (IPBES secretariat, 2022).
5 The narrative approach is influenced by the educational philosophy of John Dewey. Our narrative research method drew on the work of B. Czarniawska, *The Uses of Narrative in Social Science Research: Handbook of Data Analysis* (Sage, 2004), pp. 649–66; S. Chase, *Narrative Inquiry: The Sage Handbook of Qualitative Research* (Sage, 2005), pp. 651–79; and D. E. Polkinghorne, "Phenomenological Research Methods", in *Existential-Phenomenological Perspectives in Psychology*, ed. by R. S. Valle and S. Halling (Plenum, 1989), pp. 41–60.
6 R. Atkinson, "The Life Story Interview", in *Qualitative Research Methods Series*, 44 (Sage Publications, 1998), p. 8.
7 Atkinson, p. 14.
8 J. A. Smith, "Thematic Analysis", in *Qualitative Psychology: A Practical Guide to Research Methods* (Sage, 2015), pp. 222–48.
9 D. Harper and A. R. Thompson, eds., *Thematic Analysis in Qualitative Research Methods in Mental Health: A Guide for Students and Practitioners* (John Wiley & Sons, 2011).
10 Harper and Thompson, pp. 222–48.
11 Our participants resided in the villages of Burans Khanda (altitude 2,196 metres); Rauto ki Valley/Beli (altitude 1,863 metres); Suwakholi (altitude 2,069 metres); Landour (altitude 2,169 metres); Almas (altitude 1,512 metres); and Dhanaulti (altitude 2,286 metres) within this state.
12 Participants included residents of Kalimpong (Latitude 260 51′, longitude 880 53′, altitude 1,250 metres; Kurseong (latitude 26.88210 N, longitude 88.27890 E, altitude 1,500 metres); Siliguri (latitude 27.72710 N, longitude 88.39520 E, altitude 140 metres); and those living on the outskirts of Siliguri village in North Bengal.
13 According to GPS data.
14 Homestays are guest houses within the living quarters of a family where guests are living as members of the family. This gives them the opportunity to be familiar with the lifestyle and culture of the place.
15 Vishwambhar Prasad Sati, *Delhi Post*, 1 March 2021, https://delhipostnews.com/author/vishwambharprasad/ [accessed 20 November 2024].
16 Sati, *Delhi Post*.
17 Sati, *Delhi Post*.
18 Sati, *Delhi Post*.
19 Lertzman, *Environmental Melancholia: Psychoanalytic Dimensions of Engagement*, p. 105.
20 Lertzman, p. 81.
21 Lertzman, p. 78.

22 Isis Brook, "The Importance of Nature. Green Spaces, and Gardens in Human Well-Being", *Ethics, Policy and Environment*, 13.3 (2020), pp. 295–312.
23 Lertzman, *Environmental Melancholia: Psychoanalytic Dimensions of Engagement*, p. 75.
24 Lertzman, p. 80.
25 A holy shrine of Hindus situated at an altitude of 10,500 feet in the northern Himalayas.
26 Brook, pp. 295–312.
27 Lertzman, *Environmental Melancholia: Psychoanalytic Dimensions of Engagement*, p. 79.
28 R. Britton, "The Oedipal Situation and the Depressive Situation", in *Clinical Lectures on Klein and Bion*, ed. by R. Anderson (Routledge, 1992), p. 103.
29 D. W. Winnicott, "The Theory of the Parent-Child Relationship", *International Journal of Psychoanalysis*, 41 (1960), pp. 585–95; and E. Bick, "The Experience of the Skin in Early Object Relations", *International Journal of Psychoanalysis*, 49 (1968), p. 484.
30 Susan. M. Koger, *The Psychology of Environmental Problems* (Psychology Press, 2010), p. 81.
31 Carolyn Merchant, *The Death of Nature: Women, Ecology and the Scientific Revolution* (Harper, 1980).
32 See Weintrobe, *Engaging with Climate Change*.
33 Anderson et al., "Conceptualizing the Diverse Values of Nature and Their Contributions to People".

Index

psychic landscape 201
psychic object 2
psychic processes, ecology
  (relationship) 63–6
psychic reality 154; contact, fear 25;
  reference 20–1
psychic retreat, existence 99
psychic time 127
psychic world, nature (role) 180–1
psychoanalysis 121–3, 180; conception
  (Freud) 111; culprit, search 14;
  nature, relationship 106–7;
  recovering psychoanalysis, need
  5–6; schism 3
psychoanalysts, method 54–5
psychoanalytical enlightenment:
  climate crisis, challenge 94–6;
  defining 93
psychoanalytic anthropology 68–70
psychoanalytic community criticism 16
psychoanalytic data, usage 54
psychoanalytic discourse, external/
  internal reality 13–16
psychoanalytic dyad 77
psychoanalytic field, extensions 63–6
psychoanalytic frame, aspect 141–5
psychoanalytic matters 131
Psychoanalytic Publishing Association,
  search engine (usage) 12
psychoanalytic query 138–41
psychoanalytic reading 69
psychoanalytic treatment, climate
  change (relationship) 16–18
psychoanalytic work, reciprocity 74–5
Psychological Roots of the Climate Crisis
  (Weintrobe) 51, 111
psychological stories 181–2
psychosis, power 45
psychotic depression, suffering 16, 81
psychotic fantasy 13–14
psychotic functioning 169
public affairs, clear-sightedness 110
public welfare 110
punishing world, retreat 26–7
pure ego ideal 84
purity, obscuring/spoiling 15

race, problem 53
racial inequality, wounds 127
racism: implicit racism 5–6;
  presence 170
Radical Hope (Lear) 163
Randall, Ro 2
reality: alternative, pressure 43;
  contested idea 51; impact 14

real object, open contact 30
reason, use 116
recapitulation, theory 106
recognition: feeling 29; reciprocal
  relationships (Hegel) 97; social
  orders, transformation 99; theory
  97–8
recovering psychoanalysis, need 5–6
recovery, observation 148
refugees, sympathy 203
refuge, place 20
regressive pull 18
rejection, experiences 19
relational affinity 120
relational turn 94
relief: dynamics, highlighting 14; sense
  155–6
"remembrance of nature in the subject"
  96, 101
reparation: description 32; Eros impulse
  25–6; feeling 27; theme 6
reparative choices 66
repressed sexuality, Freud focus 94
resistance, non-totalitarian world 171
resources, self-limited usage 60
responsibility, sense 59
Reston, James 87
revenge 40; feeling 39
Rey, Henri 32
Ribeiro, Darcy 118
rich, taxes (increase) 13
Richter, Horst-Eberhard 95
Rifkin, Jeremy 61
rivers, relationship 196–7
Roots: The Saga of an American Family
  (Haley) 155
Rose, Jacqueline 127
runaway processes 62
"running out of control" sensation 62

sacred, realm (entry) 182
sadness, feelings 28
sadomasochistic battlefield 47
sadomasochistic defences 47
sadomasochistic refuge 29
sadomasochistic transference 36
sanctuary 28
Sandler, Joseph 3, 159
sanity, foundering 31
Sas, Amati 167
Schinaia, Cosimo 63
schizo-paranoid position 169
Schweitzer, Robert D. 156
sea, plastic presence (reflections) 58;
  dream 66–8

www.ingramcontent.com/pod-product-compliance
Lightning Source LLC
Chambersburg PA
CBHW062117020426
42335CB00013B/996